In the Heart of the Amazon Forest

In the Heart of the Amazon Forest, 1859

HENRY WALTER BATES

In the Heart of the Amazon Forest

GREAT JOURNEYS

PENGUIN BOOKS

Published by the Penguin Group
Penguin Books Ltd, 80 Strand, London WC2R ORL, England
Penguin Group (USA) Inc., 375 Hudson Street, New York, New York 10014, USA
Penguin Group (Canada), 90 Eglinton Avenue East, Suite 700, Toronto, Ontario, Canada M4P 2Y3
(a division of Pearson Penguin Canada Inc.)
Penguin Ireland, 25 St Stephen's Green, Dublin 2, Ireland (a division of Penguin Books Ltd)
Penguin Group (Australia), 250 Camberwell Road, Camberwell, Victoria 3124, Australia
(a division of Pearson Australia Group Pty Ltd)
Penguin Books India Pvt Ltd, 11 Community Centre, Panchsheel Park, New Delhi – 110 017, India
Penguin Group (NZ), 67 Apollo Drive, Mairangi Bay, Auckland 1310, New Zealand
(a division of Pearson New Zealand Ltd)
Penguin Books (South Africa) (Pty) Ltd, 24 Sturdee Avenue, Rosebank, Johannesburg 2196, South Africa

Penguin Books Ltd, Registered Offices: 80 Strand, London WC2R ORL, England

www.penguin.com

The Naturalist on the River Amazons first published in 1863
This extract published in Penguin Books 2007
1

Inside-cover maps by Jeff Edwards

Typeset by Rowland Phototypesetting Ltd, Bury St Edmunds, Suffolk
Printed in England by Clays Ltd, St Ives plc

ISBN: 978-0-141-02539-1

Contents

The extraordinary eleven year residence on the Amazon of Henry Walter Bates (1850–1918) resulted in a haul of some 15,000 specimens, principally insects, well over half previously unknown to science. It also resulted in one of the most entertaining and evocative of all travel books, *The Naturalist on the River Amazons*. Journeying up the Amazon (or Amazons as it was then called), initially with Alfred Russel Wallace, Bates had an infinity of attractive adventures with snakes, cockroaches and vampire bats but also became a major scientist, returning to England to become an important supporter of Darwin and to develop the theory of 'Batesian mimicry', based on his study of Brazilian insects.

The sections here deal with part of his eventful extended stay in Ega (now called Tefé), where the river systems of the Amazon (then called on these upper reaches the River Solimoens), the Teffe/Tefé and the Japurá meet. His companion Antonio Cardozo is the seemingly infinitely obliging *Delegado* of Police in Ega. *Mameluco* is a Portuguese word for the offspring of Portuguese and Amerindian parents. *Tracajas* and *Aiyussas* are kinds of Amazonian turtle. The mid-river 'royal islands' referred to reflect Brazil's being ruled in this period by the Emperor Pedro II.

Blow-guns, Turtle-hunting and Alligators

I will now proceed to give some account of the more interesting of my shorter excursions in the neighbourhood of Ega. The incidents of the longer voyages, which occupied each several months, will be narrated in a separate chapter.

The settlement, as before described, is built on a small tract of cleared land at the lower or eastern end of the lake, six or seven miles from the main Amazons, with which the lake communicates by a narrow channel. On the opposite shore of the broad expanse stands a small village, called Nogueira, the houses of which are not visible from Ega, except on very clear days; the coast on the Nogueira side is high, and stretches away into the grey distance towards the southwest. The upper part of the river Teffe is not visited by the Ega people, on account of its extreme unhealthiness, and its barrenness in sarsaparilla and other wares. To Europeans it would seem a most surprising thing that the people of a civilised settlement, a hundred and seventy years old, should still be ignorant of the course of the river on whose banks their native place, for which they proudly claim the title of city, is situated. It would be very difficult for a private individual to explore it, as the necessary number of Indian paddlers could not be obtained. I knew only one person who had ascended the Teffe to any considerable distance, and he was not able to give me a distinct account of the river. The only tribe known to live on its banks are the Catauishis, a people who perforate their lips all round, and wear rows of slender sticks in the holes: their territory lies between the Purus and the Jurua, embracing both shores of the Teffe. A large, navigable stream, the Bararua, enters the lake from the west, about thirty miles above Ega; the breadth of the lake is much contracted a little below the mouth of this tributary, but it again expands further south, and terminates abruptly where the Teffe proper, a narrow river with a strong current, forms its head water.

The whole of the country for hundreds of miles is covered with picturesque but pathless forests, and there are only two roads along which excursions can be made by land from Ega. One is a narrow hunter's track, about two miles in length, which traverses the forest in the rear of the settlement. The other is an extremely pleasant path along the beach to the west of the town. This is practicable only in the dry season, when a flat strip of white sandy beach is exposed at the foot of the high wooded banks of the lake, covered with trees, which, as there is no underwood, form a spacious shady grove. I rambled daily, during many weeks of each successive dry season, along this delightful road. The trees, many of which are myrtles and wild Guavas, with smooth yellow stems, were in flower at this time; and the rippling waters of the lake, under the cool shade, everywhere bordered the path. The place was the resort of kingfishers, green and blue tree-creepers, purple-headed tanagers, and hummingbirds. Birds generally, however, were not numerous. Every tree was tenanted by Cicadas, the reedy notes of which produced that loud, jarring, insect music which is the general accompaniment of a woodland ramble in a hot climate. One species was very handsome, having wings adorned with patches of bright green and scarlet. It was very common – sometimes three or four tenanting a single tree, clinging as usual to the branches. On approaching a tree thus peopled, a number of little jets of a clear liquid would be seen squirted from aloft. I have often received the well-directed discharge full on my face; but the liquid is harmless, having a sweetish taste, and is ejected by the insect from the anus, probably in self-defence, or from fear. The number and variety of gaily-tinted butterflies, sporting about in this grove on sunny days, were so great that the bright moving flakes of colour gave quite a character to the physiognomy of the place. It was impossible to walk far without disturbing flocks of them from the damp sand at the edge of the water, where they congregated to imbibe the moisture. They were of almost all colours, sizes, and shapes: I noticed here altogether eighty species, belonging to twenty-two different genera. It is a singular fact that, with very few exceptions, all the individuals of these various species thus sporting in sunny places were of the male sex; their partners, which are much more soberly dressed and immensely less numerous than the males, being confined to the shades of the woods. Every afternoon, as the sun was getting low, I used to notice these gaudy sunshine-loving swains trooping off to the forest, where I suppose they would find their sweethearts and wives. The most abundant, next to the very common sulphur-yellow and orange-coloured kinds, were about a dozen species of Eunica, which are of large size, and are conspicuous from their liveries of glossy dark-blue and purple. A superbly-adorned creature, the Callithea Markii, having wings of a thick texture, coloured sapphire-blue and orange, was only an occasional visitor. On certain days, when the weather was very calm, two small gilded-green species (Symmachia Trochilus and Colubris) literally swarmed on the sands, their glittering wings lying wide open on the flat surface. The beach terminates, eight miles beyond Ega, at the mouth of a rivulet; the character of the coast then changes, the river banks being masked by a line of low islets amid a labyrinth of channels.

In all other directions my very numerous excursions were by water; the most interesting of those made in the immediate neighbourhood were to the houses of Indians on the banks of retired creeks – an account of one of these trips will suffice.

On the 23rd of May, 1850, I visited, in company with Antonio Cardozo, the Delegado, a family of the Passe tribe, who live near the head waters of the Igarape, which flows from the south into the Teffe, entering it at Ega. The creek is more than a quarter of a mile broad near the town, but a few miles inland it gradually contracts, until it becomes a mere rivulet flowing through a broad dell in the forest. When the river rises it fills this dell; the trunks of the lofty trees then stand many feet deep in the water, and small canoes are able to travel the distance of a day's journey under the shade, regular paths or alleys being cut through the branches and lower trees. This is the general character of the country of the Upper Amazons; a land of small elevation and abruptly undulated, the hollows forming narrow valleys in the dry months, and deep navigable creeks in the wet months. In retired nooks on the margins of these shady rivulets, a few families or small hordes of aborigines still linger in nearly their primitive state, the relicts of their once numerous tribes. The family we intended to visit on this trip was that of Pedro-uassu (Peter the Great, or Tall Peter), an old chieftain or Tushaua of the Passes.

We set out at sunrise, in a small igarite, manned by six young Indian paddlers. After travelling about three miles along the broad portion of the creek – which, being surrounded by woods, had the appearance of a large pool – we came to a part where our course seemed to be stopped by an impenetrable hedge of trees and bushes. We were some time before finding the entrance, but when fairly within the shades, a remarkable scene presented itself. It was my first introduction to these singular waterpaths. A narrow and tolerably straight alley stretched away for a long distance before us; on each side were the tops of bushes and young trees, forming a kind of border to the path, and the trunks of the tall forest trees rose at irregular intervals from the water, their crowns interlocking far over our heads, and forming a thick shade. Slender air roots hung down in clusters, and looping sipos dangled from the lower branches; bunches of grass, tillandsiae, and ferns sat in the forks of the larger boughs, and the trunks of trees near the water had adhering to them round dried masses of freshwater sponges. There was no current perceptible, and the water was stained of a dark olive-brown hue, but the submerged stems could be seen through it to a great depth. We travelled at good speed for three hours along this shady road – the distance of Pedro's house from Ega being about twenty miles. When the paddlers rested for a time, the stillness and gloom of the place became almost painful: our voices waked dull echoes as we conversed, and the noise made by fishes occasionally whipping the surface of the water was quite startling. A cool, moist, clammy air pervaded the sunless shade.

The breadth of the wooded valley, at the commencement, is probably more than half a mile, and there is a tolerably clear view for a considerable distance on each side of the waterpath through the irregular colonnade of trees; other paths also, in this part, branch off right and left from the principal road, leading to the scattered houses of Indians on the mainland. The dell contracts gradually towards the head of the rivulet, and the forest then becomes denser; the waterpath also diminishes in width, and becomes more winding, on account of the closer growth of the trees. The boughs of some are stretched forth at no great height over one's head, and are seen to be loaded with epiphytes; one orchid I noticed particularly, on account of its bright yellow flowers growing at the end of flower-stems several feet long. Some of the trunks, especially those of palms, close beneath their crowns, were clothed with a thick mass of glossy shield-shaped Pothos plants, mingled with ferns. Arrived at this part we were, in fact, in the heart of the virgin forest. We heard no noises of animals in the trees, and saw only one bird, the sky-blue chatterer, sitting alone on a high branch. For some distance the lower vegetation was so dense that the road runs under an arcade of foliage, the branches having been cut away only sufficiently to admit of the passage of a small canoe. These thickets are formed chiefly of bamboos, whose slender foliage and curving stems arrange themselves in elegant, feathery bowers; but other social plants – slender green climbers with tendrils so eager in aspiring to grasp the higher boughs that they seem to be endowed almost with animal energy, and certain low trees having large elegantly-veined leaves – contribute also to the jungly masses. Occasionally we came upon an uprooted tree lying across the path, its voluminous crown still held up by thick cables of sipo, connecting it with standing trees; a wide circuit had to be made in these cases, and it was sometimes difficult to find the right path again.

At length we arrived at our journey's end. We were then in a very dense and gloomy part of the forest – we could see, however, the dry land on both sides of the creek, and to our right a small sunny opening appeared, the landing place to the native dwellings. The water was deep close to the bank, and a clean pathway ascended from the shady port to the buildings, which were about a furlong distant. My friend Cardozo was godfather to a grandchild of Pedro-uassu, whose daughter had married an Indian settled in Ega. He had sent word to the old man that he intended to visit him: we were therefore expected.

As we landed, Pedro-uassu himself came down to the port to receive us, our arrival having been announced by the barking of dogs. He was a tall and thin old man, with a serious, but benignant expression of countenance, and a manner much freer from shyness and distrust than is usual with Indians. He was clad in a shirt of coarse cotton cloth, dyed with murishi, and trousers of the same material turned up to the knee. His features were sharply delineated – more so than in any Indian face I had yet seen; the lips thin and the nose rather high and compressed. A large, square, blue-black tattooed patch occupied the middle of his face, which, as well as the other exposed parts of his body, was of a light reddish-tan colour, instead of the usual coppery-brown hue. He walked with an upright, slow gait, and on reaching us saluted Cardozo with the air of a man who wished it to be understood that he was dealing with an equal. My friend introduced me, and I was welcomed in the same grave, ceremonious manner. He seemed to have many questions to ask, but they were chiefly about Senora Felippa, Cardozo's Indian housekeeper at Ega, and were purely complimentary. This studied politeness is quite natural to Indians of the advanced agricultural tribes. The language used was Tupi – I heard no other spoken all the day. It must be borne in mind that Pedro-uassu had never had much intercourse with whites; he was, although baptised, a primitive Indian who had always lived in retirement, the ceremony of baptism having been gone through, as it generally is by the aborigines, simply from a wish to stand well with the whites.

Arrived at the house, we were welcomed by Pedro's wife: a thin, wrinkled, active old squaw, tattooed in precisely the same way as her husband. She also had sharp features, but her manner was more cordial and quicker than that of her husband: she talked much, and with great inflection of voice; while the tones of the old man were rather drawling and querulous. Her clothing was a long petticoat of thick cotton cloth, and a very short chemise, not reaching to her waist. I was rather surprised to find the grounds around the establishment in neater order than in any sitio, even of civilised people, I had yet seen on the Upper Amazons; the stock of utensils and household goods of all sorts was larger, and the evidences of regular industry and plenty more numerous than one usually perceives in the farms of civilised Indians and whites. The buildings were of the same construction as those of the humbler settlers in all other parts of the country. The family lived in a large, oblong, open shed built under the shade of trees. Two smaller buildings, detached from the shed and having mud-walls with low doorways, contained apparently the sleeping apartments of different members of the large household. A small mill for grinding sugar-cane, having two cylinders of hard notched wood, wooden troughs, and kettles for boiling the guarapa (cane juice) to make treacle, stood under a separate shed, and near it was a large enclosed mud-house for poultry. There was another hut and shed a short distance off, inhabited by a family dependent on

Pedro, and a narrow pathway through the luxuriant woods led to more dwellings of the same kind. There was an abundance of fruit trees around the place, including the never-failing banana, with its long, broad, soft green leaf-blades, and groups of full-grown Pupunhas, or peach palms. There was also a large number of cotton and coffee trees. Among the utensils I noticed baskets of different shapes, made of flattened maranta stalks, and dyed various colours. The making of these is an original art of the Passes, but I believe it is also practised by other tribes, for I saw several in the houses of semi-civilised Indians on the Tapajos.

There were only three persons in the house besides the old couple, the rest of the people being absent; several came in, however, in the course of the day. One was a daughter of Pedro's, who had an oval tattooed spot over her mouth; the second was a young grandson; and the third the son-in-law from Ega, Cardozo's compadre. The old woman was occupied, when we entered, in distilling spirits from cara, an edible root similar to the potato, by means of a clay still, which had been manufactured by herself. The liquor had a reddish tint, but not a very agreeable flavour. A cup of it, warm from the still, however, was welcome after our long journey. Cardozo liked it, emptied his cup, and replenished it in a very short time. The old lady was very talkative, and almost fussy in her desire to please her visitors. We sat in tucum hammocks, suspended between the upright posts of the shed. The young woman with the blue mouth – who, although married, was as shy as any young maiden of her race – soon became employed in scalding and plucking fowls for the dinner near the fire on the ground at the other end of the dwelling. The son-in-law, Pedro-uassu, and Cardozo now began a long conversation on the subject of their deceased wife, daughter, and comadre. [Co-mother; the term expressing the relationship of a mother to the godfather of her child.] It appeared she had died of consumption – 'tisica,' as they called it, a word adopted by the Indians from the Portuguese. The widower repeated over and over again, in nearly the same words, his account of her illness, Pedro chiming in like a chorus, and Cardozo moralising and condoling. I thought the cauim (grog) had a good deal to do with the flow of talk and warmth of feeling of all three; the widower drank and wailed until he became maundering, and finally fell asleep.

I left them talking, and took a long ramble into the forest, Pedro sending his grandson, a smiling well-behaved lad of about fourteen years of age, to show me the paths, my companion taking with him his Zarabatana, or blow-gun. This instrument is used by all the Indian tribes on the Upper Amazons. It is generally nine or ten feet long, and is made of two separate lengths of wood, each scooped out so as to form one-half of the tube. To do this with the necessary accuracy requires an enormous amount of patient labour, and considerable mechanical ability, the tools used being simply the incisor teeth of the Paca and Cutia. The two half tubes, when finished, are secured together by a very close and tight spirally-wound strapping, consisting of long flat strips of Jacitara, or the wood of the climbing palm-tree; and the whole is smeared afterwards with black wax, the production of a Melipona bee. The pipe tapers towards the muzzle, and a cup-shaped mouthpiece, made of wood, is fitted in the broad end. A full-sized Zarabatana is heavy, and can only be used by an adult Indian who has had great practice. The young lads learn to shoot with smaller and lighter tubes. When Mr Wallace and I had lessons at Barra in the use of the blow-gun, of Julio, a Juri Indian, then in the employ of Mr Hauxwell, an English bird-collector, we found it very difficult to hold steadily the long tubes. The arrows are made from the hard rind of the leaf-stalks of certain palms, thin strips being cut, and rendered as sharp as needles by scraping the ends with a knife or the tooth of an animal. They are winged with a little oval mass of samauma silk (from the seed-vessels of the silk-cotton tree, Eriodendron samauma), cotton being too heavy. The ball of samauma should fit to a nicety the bore of the blowgun; when it does so, the arrow can be propelled with such force by the breath that it makes a noise almost as loud as a pop-gun on flying from the muzzle. My little companion was armed with a quiver full of these little missiles, a small number of which, sufficient for the day's sport, were tipped with the fatal Urari poison. The quiver was an ornamental affair, the broad rim being made of highly-polished wood of a rich cherry-red colour (the Moira-piranga, or redwood of the Japura). The body was formed of neatly-plaited strips of Maranta stalks, and the belt by which it was suspended from the shoulder was decorated with cotton fringes and tassels.

We walked about two miles along a well-trodden pathway, through high caapoeira (second-growth forest). A large proportion of the trees were Melastomas, which bore a hairy yellow fruit, nearly as large and as well flavoured as our gooseberry. The season, however, was nearly over for them. The road was bordered every inch of the way by a thick bed of elegant Lycopodiums. An artificial arrangement of trees and bushes could scarcely have been made to wear so finished an appearance as this naturally decorated avenue. The path at length terminated at a plantation of mandioca, the largest I had yet seen since I left the neighbourhood of Para. There were probably ten acres of cleared land, and part of the ground was planted with Indian corn, water-melons, and sugar cane. Beyond this field there was only a faint hunter's track, leading towards the untrodden interior. My companion told me he had never heard of there being any inhabitants in that direction (the south). We crossed the forest from this place to another smaller clearing, and then walked, on our road home, through about two miles of caapoeira of various ages, the sites of old plantations. The only fruits of our ramble were a few rare insects and a Japu (Cassicus cristatus), a handsome bird with chestnut and saffron-coloured plumage, which wanders through the tree-tops in large flocks. My little companion brought this down from a height which I calculated at thirty yards. The blow-gun, however, in the hands of an expert adult Indian, can be made to propel arrows so as to kill at a distance of fifty and sixty yards. The aim is most certain when the tube is held vertically, or nearly so. It is a far more useful weapon in the forest than a gun, for the report of a firearm alarms the whole flock of birds or monkeys feeding on a tree, while the silent poisoned dart brings the animals down one by one until the sportsman has a heap of slain by his side. None but the stealthy Indian can use it effectively. The poison, which must be fresh to kill speedily, is obtained only of the Indians who live beyond the cataracts of the rivers flowing from the north, especially the Rio Negro and the Japura. Its principal ingredient is the wood of the Strychnos toxifera, a tree which does not grow in the humid forests of the river plains. A most graphic account of the Urari, and of an expedition undertaken in search of the tree in Guiana, has been given by Sir Robert Schomburgk.

When we returned to the house after mid-day, Cardozo was still sipping cauim, and now looked exceedingly merry. It was fearfully hot; the good fellow sat in his hammock with a cuya full of grog in his hands; his broad honest face all of a glow, and the perspiration streaming down his uncovered breast, the unbuttoned shirt having slipped half-way over his broad shoulders. Pedro-uassu had not drunk much; he was noted, as I afterwards learned, for his temperance. But he was standing up as I had left him two hours previous, talking to Cardozo in the same monotonous tones, the conversation apparently not having flagged all the time. I had never heard so much talking amongst

Indians. The widower was asleep; the stirring, managing old lady with her daughter were preparing dinner. This, which was ready soon after I entered, consisted of boiled fowls and rice, seasoned with large green peppers and lemon juice, and piles of new, fragrant farinha and raw bananas. It was served on plates of English manufacture on a tupe, or large plaited rush mat, such as is made by the natives pretty generally on the Amazons. Three or four other Indians, men and women of middle age, now made their appearance, and joined in the meal. We all sat round on the floor: the women, according to custom, not eating until after the men had done. Before sitting down, our host apologised in his usual quiet, courteous manner for not having knives and forks; Cardozo and I ate by the aid of wooden spoons, the Indians using their fingers. The old man waited until we were all served before he himself commenced. At the end of the meal, one of the women brought us water in a painted clay basin of Indian manufacture, and a clean coarse cotton napkin, that we might wash our hands.

The horde of Passes of which Pedro-uassu was Tushaua or chieftain, was at this time reduced to a very small number of individuals. The disease mentioned in the last chapter had for several generations made great havoc among them; many had also entered the service of whites at Ega, and, of late years, intermarriages with whites, half-castes, and civilised Indians had been frequent. The old man bewailed the fate of his race to Cardozo with tears in his eyes. 'The people of my nation,' he said, 'have always been good friends to the

Cariwas (whites), but before my grandchildren are old like me the name of Passe will be forgotten.' In so far as the Passes have amalgamated with European immigrants or their descendants, and become civilised Brazilian citizens, there can scarcely be ground for lamenting their extinction as a nation; but it fills one with regret to learn how many die prematurely of a disease which seems to arise on their simply breathing the same air as the whites. The original territory of the tribe must have been of large extent, for Passes are said to have been found by the early Portuguese colonists on the Rio Negro; an ancient settlement on that river, Barcellos, having been peopled by them when it was first established; and they formed also part of the original population of Fonteboa on the Solimoens. Their hordes were therefore, spread over a region 400 miles in length from east to west. It is probable, however, that they have been confounded by the colonists with other neighbouring tribes who tattoo their faces in a similar manner. The extinct tribe of Yurimauas, or Sorimoas, from which the river Solimoens derives its name, according to traditions extant at Ega, resembled the Passes in their slender figures and friendly disposition. These tribes (with others lying between them) peopled the banks of the main river and its by-streams from the mouth of the Rio Negro to Peru. True Passes existed in their primitive state on the banks of the Issa, 240 miles to the west of Ega, within the memory of living persons. The only large body of them now extant are located on the Japura, at a place distant about 150 miles from Ega: the population of this horde, however, does not exceed, from what I could learn, 300 or 400 persons. I think it probable that the lower part of the Japura and its extensive delta lands formed the original home of this gentle tribe of Indians.

The Passes are always spoken of in this country as the most advanced of all the Indian nations in the Amazons region. Under what influences this tribe has become so strongly modified in mental, social, and bodily features it is hard to divine. The industrious habits, fidelity, and mildness of disposition of the Passes, their docility and, it may be added, their personal beauty, especially of the children and women, made them from the first very attractive to the Portuguese colonists. They were, consequently, enticed in great number from their villages and brought to Barra and other settlements of the whites. The wives of governors and military officers from Europe were always eager to obtain children for domestic servants; the girls being taught to sew, cook, weave hammocks, manufacture pillow-lace, and so forth. They have been generally treated with kindness, especially by the educated families in the settlements. It is pleasant to have to record that I never heard of a deed of violence perpetrated, on the one side or the other, in the dealings between European settlers and this noble tribe of savages.

Very little is known of the original customs of the Passes. The mode of life of our host Pedro-uassu did not differ much from that of the civilised Mamelucos; but he and his people showed a greater industry, and were more open, cheerful, and generous in their dealings than many half-castes. The authority of Pedro, like that of the Tushauas, generally was exercised in a mild manner. These chieftains appear able to command the services of their subjects, since they furnish men to the Brazilian authorities when requested; but none of them, even those of the most advanced tribes, appear to make use of this authority for the accumulation of property – the service being exacted chiefly in time of war. Had the ambition of the chiefs of some of these industrious tribes been turned to the acquisition of wealth, probably we should have seen indigenous civilised nations in the heart of South America similar to those found on the Andes of Peru and Mexico. It is very probable that the Passes adopted from the first to some extent the manners of the whites. Ribeiro, a Portuguese official who travelled in these regions in 1774–5, and wrote an account of his journey, relates that they buried their dead in large earthenware vessels (a custom still observed among other tribes on the Upper Amazons), and that, as to their marriages, the young men earned their brides by valiant deeds in war. He also states that they possessed a cosmogony in which the belief that the sun was a fixed body, with the earth revolving around it, was a prominent feature. He says, moreover, that they believed in a Creator of all things; a future state of rewards and punishments, and so forth. These notions are so far in advance of the ideas of all other tribes of Indians, and so little likely to have been conceived and perfected by a people having no written language or leisured class, that we must suppose them to have been derived by the docile

Passes from some early missionary or traveller. I never found that the Passes had more curiosity or activity of intellect than other Indians. No trace of a belief in a future state exists amongst Indians who have not had much intercourse with the civilised settlers, and even amongst those who have it is only a few of the more gifted individuals who show any curiosity on the subject. Their sluggish minds seem unable to conceive or feel the want of a theory of the soul, and of the relations of man to the rest of Nature or to the Creator. But is it not so with totally uneducated and isolated people even in the most highly civilised parts of the world? The good qualities of the Passes belong to the moral part of the character: they lead a contented, unambitious, and friendly life, a quiet, domestic, orderly existence, varied by occasional drinking bouts and summer excursions. They are not so shrewd, energetic, and masterful as the Mundurucus, but they are more easily taught, because their disposition is more yielding than that of the Mundurucus or any other tribe.

We started on our return to Ega at half-past four o'clock in the afternoon. Our generous entertainers loaded us with presents. There was scarcely room for us to sit in the canoe, as they had sent down ten large bundles of sugar-cane, four baskets of farinha, three cedar planks, a small hamper of coffee, and two heavy bunches of bananas. After we were embarked, the old lady came with a parting gift for me – a huge bowl of smoking hot banana porridge. I was to eat it on the road 'to keep my stomach warm.' Both stood on the bank as we pushed off, and gave us their adios: 'Ikudna

Tupana eirum' (Go with God) – a form of salutation taught by the old Jesuit missionaries. We had a most uncomfortable passage, for Cardozo was quite tipsy, and had not attended to the loading of the boat. The cargo had been placed too far forward, and to make matters worse, my heavy friend obstinately insisted on sitting astride on the top of the pile, instead of taking his place near the stern, singing from his perch a most indecent love-song, and disregarding the inconvenience of having to bend down almost every minute to pass under the boughs of hanging sipos as we sped rapidly along. The canoe leaked but not, at first, alarmingly. Long before sunset, darkness began to close in under those gloomy shades, and our steersman could not avoid now and then running the boat into the thicket. The first time this happened a piece was broken off the square prow (rodella); the second time we got squeezed between two trees. A short time after this latter accident, being seated near the stern with my feet on the bottom of the boat, I felt rather suddenly the cold water above my ankles. A few minutes more and we should have sunk, for a seam had been opened forward under the pile of sugar-cane. Two of us began to bale, and by the most strenuous efforts managed to keep afloat without throwing overboard our cargo. The Indians were obliged to paddle with extreme slowness to avoid shipping water, as the edge of our prow was nearly level with the surface; but Cardozo was now persuaded to change his seat. The sun set, the quick twilight passed, and the moon soon after began to glimmer through the thick canopy of foliage. The prospect of being swamped in this hideous solitude was by no means pleasant, although I calculated on the chance of swimming to a tree and finding a nice snug place in the fork of some large bough wherein to pass the night.

At length, after four hours' tedious progress, we suddenly emerged on the open stream where the moonlight glittered in broad sheets on the gently rippling waters. A little extra care was now required in paddling. The Indians plied their strokes with the greatest nicety; the lights of Ega (the oil lamps in the houses) soon appeared beyond the black wall of forest, and in a short time we leapt safely ashore.

A few months after the excursion just narrated, I accompanied Cardozo in many wanderings on the Solimoens, during which he visited the praias (sand-islands), the turtle pools in the forests, and the by-streams and lakes of the great desert river. His object was mainly to superintend the business of digging up turtle eggs on the sandbanks, having been elected commandante for the year by the municipal council of Ega, of the 'praia real' (royal sand-island) of Shimuni, the one lying nearest to Ega. There are four of these royal praias within the Ega district (a distance of 150 miles from the town), all of which are visited annually by the Ega people for the purpose of collecting eggs and extracting oil from their yolks. Each has its commander, whose business is to make arrangements for securing to every inhabitant an equal chance in the egg harvest by placing sentinels to protect the turtles whilst laying, and so forth. The pregnant turtles descend from the interior pools to the main river in July and August, before the outlets dry up, and then seek in countless swarms their favourite sand islands; for it is only a few praias that are selected by them out of the great number existing. The young animals remain in the pools throughout the dry season. These breeding places of turtles then lie twenty to thirty or more feet above the level of the river, and are accessible only by cutting roads through the dense forest.

We left Ega on our first trip to visit the sentinels while the turtles were yet laying, on the 26th of September. Our canoe was a stoutly built igarite, arranged for ten paddlers, and having a large arched toldo at the stern under which three persons could sleep pretty comfortably. Emerging from the Teffe we descended rapidly on the swift current of the Solimoens to the south-eastern or lower end of the large wooded island of Baria, which here divides the river into two great channels. We then paddled across to Shimuni, which lies in the middle of the northeasterly channel, reaching the commencement of the praia an hour before sunset. The island proper is about three miles long and half a mile broad: the forest with which it is covered rises to an immense and uniform height, and presents all round a compact, impervious front. Here and there a singular tree, called Pao mulatto (mulatto wood), with polished dark-green trunk, rose conspicuously among the mass of vegetation. The sandbank, which lies at the upper end of the island, extends several miles and presents an irregular, and in some parts, strongly-waved surface, with deep hollows and ridges. When upon it, one feels as though treading an almost boundless field of sand, for towards the southeast, where no forest line terminates the view, the white, rolling plain stretches away to the horizon. The north-easterly channel of the river lying between the sands and the further shore of the river is at least two miles in breadth; the middle one, between the two islands, Shimuni and Baria, is not much less than a mile.

We found the two sentinels lodged in a corner of the praia, where it commences at the foot of the towering forest wall of the island, having built for themselves a little rancho with poles and palm-leaves. Great precautions are obliged to be taken to avoid disturbing the sensitive turtles, who, previous to crawling ashore to lay, assemble in great shoals off the sandbank. The men, during this time, take care not to show themselves and warn off any fisherman who wishes to pass near the place. Their fires are made in a deep hollow near the borders of the forest, so that the smoke may not be visible. The passage of a boat through the shallow waters where the animals are congregated, or the sight of a man or a fire on the sandbank, would prevent the turtles from leaving the water that night to lay their eggs, and if the causes of alarm were repeated once or twice, they would forsake the praia for some other quieter place. Soon after we arrived, our men were sent with the net to catch a supply of fish for supper. In half an hour, four or five large basketful of Acari were brought in. The sun set soon after our meal was cooked; we were then obliged to extinguish the fire and remove our supper materials to the sleeping ground, a spit of sand about a mile off – this course being necessary on account of the mosquitoes which swarm at night on the borders of the forest.

One of the sentinels was a taciturn, morose-looking, but sober and honest Indian, named Daniel; the other was a noted character of Ega, a little wiry Mameluco, named Carepira (Fish-hawk) – known for his waggery, propensity for strong drink, and indebtedness to Ega traders. Both were intrepid canoemen and huntsmen, and both perfectly at home anywhere in these fearful wastes of forest and water. Carepira had his son with him – a quiet little lad of about nine years of age. These men in a few minutes constructed a small shed with four upright poles and leaves of the arrow-grass, under which Cardozo and I slung our hammocks. We did not go to sleep, however, until after midnight – for when supper was over, we lay about on the sand with a flask of rum in our midst and whiled away the still hours in listening to Carepira's stories.

I rose from my hammock by daylight, shivering with cold; a praia, on account of the great radiation of heat in the night from the sand, being towards the dawn the coldest place that can be found in this climate. Cardozo and the men were already up watching the turtles. The sentinels had erected for this purpose a stage about fifty feet high, on a tall tree near their station, the ascent to which was by a roughly-made ladder of woody lianas. They are enabled, by observing the turtles from this watchtower, to ascertain the date of successive deposits of eggs, and thus guide the commandante in fixing the time for the general invitation to the Ega people. The turtles lay their eggs by night, leaving the water when nothing disturbs them, in vast crowds, and crawling to the central and highest part of the praia. These places are, of course, the last to go under water when, in unusually wet seasons, the river rises before the eggs are hatched by the heat of the sand. One could almost believe from this that the animals used forethought in choosing a place; but it is simply one of those many instances in animals where unconscious habit has the same result as conscious prevision. The hours between midnight and dawn are the busiest. The turtles excavate with their broad, webbed paws, deep holes in the fine sand – the first comer, in each case, making a pit about three feet deep, laying its eggs (about 120 in number) and covering them with sand; the next making its deposit at the top of that of its predecessor, and so on until every pit is full. The whole body of turtles frequenting a praia does not finish laying in less than fourteen or fifteen days, even when there is no interruption. When all have done, the area (called by the Brazilians taboleiro) over which they have excavated is distinguishable from the rest of the praia only by signs of the sand having been a little disturbed.

On rising, I went to join my friends. Few recollections of my Amazonian rambles are more vivid and agreeable than that of my walk over the white sea of sand on this cool morning. The sky was cloudless; the just-risen sun was hidden behind the dark mass of woods on Shimuni, but the long line of forest to the west, on Baria, with its plumy decorations of palms, was lighted up with his yellow, horizontal rays. A faint chorus of singing birds reached the ears from across the water, and flocks of gulls and plovers were crying plaintively over the swelling banks of the praia, where their eggs lay in nests made in little hollows of the sand. Tracks of stray turtles were visible on the smooth white surface of the praia. The animals which thus wander from the main body are lawful prizes of the sentinels; they had caught in this way two before sun-rise, one of which we had for dinner. In my walk I disturbed several pairs of the chocolate and drab-coloured wild-goose (Anser jubatus) which set off to run along the edge of the water. The enjoyment one feels in rambling over these free, open spaces, is no doubt enhanced by the novelty of the scene, the change being very great from the monotonous landscape of forest which everywhere else presents itself.

On arriving at the edge of the forest I mounted the sentinel's stage, just in time to see the turtles retreating to the water on the opposite side of the sand-bank, after having laid their eggs. The sight was well worth the trouble of ascending the shaky ladder. They were about a mile off, but the surface of the sands was blackened with the multitudes which were waddling towards the river; the margin of the praia was rather steep, and they all seemed to tumble head first down the declivity into the water.

I spent the morning of the 27th collecting insects in the woods of Shimuni; and assisted my friend in the afternoon to beat a large pool for Tracajas – Cardozo wishing to obtain a supply for his table at home. The pool was nearly a mile long, and lay on one side of the island between the forest and the sand-bank. The sands are heaped up very curiously around the margins of these isolated sheets of water; in the present case they formed a steeply-inclined bank, from five to eight feet in height. What may be the cause of this formation I cannot imagine. The pools always contain a quantity of imprisoned fish, turtles, Tracajas, and Aiyussas. The turtles and Aiyussas crawl out voluntarily in the course of a few days, and escape to the main river, but the Tracajas remain and become an easy prey to the natives. The ordinary mode of obtaining them is to whip the water in every part with rods for several hours during the day; this treatment having the effect of driving the animals out. They wait, however, until the night following the beating before making their exit. Our Indians were occupied for many hours in this work, and when night came they and the sentinels were placed at intervals along the edge of the water to be ready to capture the runaways. Cardozo and I, after supper, went and took our station at one end of the pool.

We did not succeed, after all our trouble, in getting many Tracajas. This was partly owing to the intense darkness of the night, and partly, doubtless, to the sentinels having already nearly exhausted the pool, notwithstanding their declarations to the contrary. In waiting for the animals, it was necessary to keep silence – not a pleasant way of passing the night . . . speaking only in whispers, and being without fire in a place liable to be visited by a prowling jaguar. Cardozo and I sat on a sandy slope with our loaded guns by our side, but it was so dark we could scarcely see each other.

Towards midnight a storm began to gather around us. The faint wind which had breathed from over the water since the sun went down, ceased. Thick clouds piled themselves up, until every star was obscured, and gleams of watery lightning began to play in the midst of the black masses. I hinted to Cardozo that I thought we had now had enough of watching, and suggested a cigarette. Just then a quick pattering movement was heard on the sands, and grasping our guns, we both started to our feet. Whatever it might have been it seemed to pass by, and a few moments afterwards a dark body appeared to be moving in another direction on the opposite slope of the sandy ravine where we lay. We prepared to fire, but luckily took the precaution of first shouting 'Quem vai la?' (Who goes there?) It turned out to be the taciturn sentinel, Daniel, who asked us mildly whether we had heard a 'raposa' pass our way. The raposa is a kind of wild dog, with very long tapering muzzle, and black and white speckled hair. Daniel could distinguish all kinds of animals in the dark by their footsteps. It now began to thunder, and our position was getting very uncomfortable. Daniel had not seen anything of the other Indians, and thought it was useless waiting any longer for Tracajas; we therefore sent him to call in the whole party, and made off ourselves, as quickly as we could, for the canoe. The rest of the night was passed most miserably; as indeed were very many of my nights on the Solimoens. A furious squall burst upon us; the wind blew away the cloths and mats we had fixed up at the ends of the arched awning of the canoe to shelter ourselves, and the rain beat right through our sleeping-place. There we lay, Cardozo and I, huddled together, and wet through, waiting for the morning.

A cup of strong and hot coffee put us to rights at sunrise, but the rain was still coming down, having changed to a steady drizzle. Our men were all returned from the pool, having taken only four Tracajas. The business which had brought Cardozo hither being now finished, we set out to return to Ega, leaving the sentinels once more to their solitude on the sands. Our return route was by the rarely frequented north-easterly channel of the Solimoens, through which flows part of the waters of its great tributary stream, the Japura. We travelled for five hours along the desolate, broken, timber-strewn shore of Baria. The channel is of immense breadth, the opposite coast being visible only as a long, low line of forest. At three o'clock in the afternoon we doubled the upper end of the island, and then crossed towards the mouth of the Teffe by a broad transverse channel running between Baria and another island called Quanaru. There is a small sand-bank at the north-westerly point of Baria, called Jacare; we stayed here to dine and afterwards fished with the net. A fine rain was still falling, and we had capital sport – in three hauls taking more fish than our canoe would conveniently hold. They were of two kinds only, the Surubim and the Piraepieua (species of Pimelodus), very handsome fishes, four feet in length, with flat spoon-shaped heads, and prettily-spotted and striped skins.

On our way from Jacare to the mouth of the Teffe

we had a little adventure with a black tiger or jaguar. We were paddling rapidly past a long beach of dried mud, when the Indians became suddenly excited, shouting 'Ecui Jauarete; Jauaripixuna!' (Behold the jaguar, the black jaguar!) Looking ahead we saw the animal quietly drinking at the water's edge. Cardozo ordered the steersman at once to put us ashore. By the time we were landed the tiger had seen us, and was retracing his steps towards the forest. On the spur of the moment, and without thinking of what we were doing, we took our guns (mine was a double-barrel, with one charge of B B and one of dust-shot) and gave chase. The animal increased his speed, and reaching the forest border, dived into the dense mass of broad-leaved grass which formed its frontage. We peeped through the gap he had made, but, our courage being by this time cooled, we did not think it wise to go into the thicket after him. The black tiger appears to be more abundant than the spotted form of jaguar in the neighbourhood of Ega. The most certain method of finding it is to hunt assisted by a string of Indians shouting and driving the game before them in the narrow restingas or strips of dry land in the forest, which are isolated by the flooding of their neighbourhood in the wet season. We reached Ega by eight o'clock that night.

On the 6th of October we left Ega on a second excursion; the principal object of Cardozo being, this time, to search certain pools in the forest for young turtles. The exact situation of these hidden sheets of water is known only to a few practised huntsmen; we took one of these men with us from Ega, a Mameluco named Pedro, and on our way called at Shimuni for Daniel to serve as an additional guide. We started from the praia at sunrise on the 7th in two canoes containing twenty-three persons, nineteen of whom were Indians. The morning was cloudy and cool, and a fresh wind blew from down river, against which we had to struggle with all the force of our paddles, aided by the current; the boats were tossed about most disagreeably, and shipped a great deal of water. On passing the lower end of Shimuni, a long reach of the river was before us, undivided by islands – a magnificent expanse of water stretching away to the southeast. The country on the left bank is not, however, terra firma, but a portion of the alluvial land which forms the extensive and complex delta region of the Japura. It is flooded every year at the time of high water, and is traversed by many narrow and deep channels which serve as outlets to the Japura, or at least, are connected with that river by means of the interior water-system of the Cupiyo. This inhospitable tract of country extends for several hundred miles, and contains in its midst an endless number of pools and lakes tenanted by multitudes of turtles, fishes, alligators, and water serpents. Our destination was a point on this coast situated about twenty miles below Shimuni, and a short distance from the mouth of the Anana, one of the channels just alluded to as connected with the Japura. After travelling for three hours in midstream we steered for the land, and brought to under a steeply-inclined bank of crumbly earth, shaped into a succession of steps or terraces, marking the various halts which the waters of the river make in the course of subsidence. The coast line was nearly straight for many miles, and the bank averaged about thirty feet in height above the present level of the river: at the top rose the unbroken hedge of forest. No one could have divined that pools of water existed on that elevated land. A narrow level space extended at the foot of the bank. On landing the first business was to get breakfast. While a couple of Indian lads were employed in making the fire, roasting the fish, and boiling the coffee, the rest of the party mounted the bank, and with their long hunting knives commenced cutting a path through the forest; the pool, called the Aningal, being about half a mile distant. After breakfast, a great number of short poles were cut and were laid crosswise on the path, and then three light montarias which we had brought with us were dragged up the bank by lianas, and rolled away to be embarked on the pool. A large net, seventy yards in length, was then disembarked and carried to the place. The work was done very speedily, and when Cardozo and I went to the spot at eleven o'clock, we found some of the older Indians, including Pedro and Daniel, had begun their sport. They were mounted on little stages called moutas, made of poles and cross-pieces of wood secured with lianas, and were shooting the turtles as they came near the surface, with bows and arrows. The Indians seemed to think that netting the animals, as Cardozo proposed doing, was not lawful sport, and wished first to have an hour or two's old-fashioned practice with their weapons.

The pool covered an area of about four or five acres, and was closely hemmed in by the forest, which in picturesque variety and grouping of trees and foliage exceeded almost everything I had yet witnessed. The margins for some distance were swampy, and covered with large tufts of a fine grass called Matupa. These tufts in many places were overrun with ferns, and exterior to them a crowded row of arborescent arums, growing to a height of fifteen or twenty feet, formed a green palisade. Around the whole stood the taller forest trees; palmate-leaved Cecropiae, slender Assai palms, thirty feet high, with their thin feathery heads crowning the gently-curving, smooth stems; small fan-leaved palms; and as a background to all these airy shapes, lay the voluminous masses of ordinary forest trees, with garlands, festoons, and streamers of leafy climbers hanging from their branches. The pool was nowhere more than five feet deep, one foot of which was not water, but extremely fine and soft mud.

Cardozo and I spent an hour paddling about. I was astonished at the skill which the Indians display in shooting turtles. They did not wait for their coming to the surface to breathe, but watched for the slight movements in the water, which revealed their presence underneath. These little tracks on the water are called the Siriri; the instant one was perceived an arrow flew from the bow of the nearest man, and never failed to pierce the shell of the submerged animal. When the turtle was very distant, of course the aim had to be taken at a considerable elevation, but the marksmen preferred a longish range, because the arrow then fell more perpendicularly on the shell and entered it more deeply.

The arrow used in turtle shooting has a strong lancet-shaped steel point, fitted into a peg which enters the tip of the shaft. The peg is secured to the shaft by twine made of the fibres of pineapple leaves, the twine being some thirty or forty yards in length, and neatly wound round the body of the arrow. When the missile enters the shell, the peg drops out, and the pierced animal descends with it towards the bottom, leaving the shaft floating on the surface. This being done, the sportsman paddles in his montaria to the place, and gently draws the animal by the twine, humouring it by giving it the rein when it plunges, until it is brought again near the surface, when he strikes it with a second arrow. With the increased hold given by the two cords he has then no difficulty in landing his game.

By mid-day the men had shot about a score of nearly full-grown turtles. Cardozo then gave orders to spread the net. The spongy, swampy nature of the banks made it impossible to work the net so as to draw the booty ashore; another method was therefore adopted. The net was taken by two Indians and extended in a curve at one extremity of the oval-shaped pool, holding it when they had done so by the perpendicular rods fixed at each end; its breadth was about equal to the depth of the water, its shotted side therefore rested on the bottom, while the floats buoyed it up on the surface, so that the whole, when the ends were brought together, would form a complete trap. The rest of the party then spread themselves around the swamp at the opposite end of the pool and began to beat, with stout poles, the thick tufts of Matupa, in order to drive the turtles towards the middle. This was continued for an hour or more, the beaters gradually drawing nearer to each other, and driving the host of animals before them; the number of little snouts constantly popping above the surface of the water showing that all was going on well. When they neared the net the men moved more quickly, shouting and beating with great vigour. The ends of the net were then seized by several strong hands and dragged suddenly forwards, bringing them at the same time together, so as to enclose all the booty in a circle. Every man now leapt into the enclosure, the boats were brought up, and the turtles easily captured by the hand and tossed into them. I jumped in along with the rest, although I had just before made the discovery that the pool abounded in ugly, red, four-angled leeches, having seen several of these delectable animals, which sometimes fasten on the legs of fishermen, although they did not, on this day, trouble us, working their way through cracks in the bottom of our montaria. Cardozo, who remained with the boats, could not turn the animals on their backs fast enough, so that a great many clambered out and got free again. However, three boat-loads, or about eighty, were secured in about twenty minutes. They were then taken ashore, and each one secured by the men tying the legs with thongs of bast.

When the canoes had been twice filled, we desisted, after a very hard day's work. Nearly all the animals were young ones, chiefly, according to the statement of Pedro, from three to ten years of age; they varied from six to eighteen inches in length, and were very fat. Cardozo and I lived almost exclusively on them for several months afterwards. Roasted in the shell they form a most appetising dish. These younger turtles never migrate with their elders on the sinking of the waters, but remain in the tepid pools, fattening on fallen fruits, and, according to the natives, on the fine nutritious mud. We captured a few full-grown mother-turtles, which were known at once by the horny skin of their breast-plates being worn, telling of their having crawled on the sands to lay eggs the previous year. They had evidently made a mistake in not leaving the pool at the proper time, for they were full of eggs, which, we were told, they would, before the season was over, scatter in despair over the swamp. We also found several male turtles, or Capitaris, as they are called by the natives. These are immensely less numerous than the females, and are distinguishable by their much smaller size, more circular shape, and the greater length and thickness of their tails. Their flesh is considered unwholesome, especially to sick people having external signs of inflammation. All diseases in these parts, as well as their remedies and all articles of food, are classed by the inhabitants as 'hot' and 'cold,' and the meat of the Capitari is settled by unanimous consent as belonging to the 'hot' list.

We dined on the banks of the river a little before sunset. The mosquitoes then began to be troublesome, and finding it would be impossible to sleep here, we all embarked and crossed the river to a sand-bank, about three miles distant, where we passed the night. Cardozo and I slept in our hammocks slung between upright poles, the rest stretching themselves on the sand round a large fire. We lay awake conversing until past midnight. It was a real pleasure to listen to the stories told by one of the older men, they were given with so much spirit. The tales always related to struggles with some intractable animal: jaguar, manatee, or alligator. Many interjections and expressive gestures were used, and at the end came a sudden 'Pa! terra!' when the animal was vanquished by a shot or a blow. Many mysterious tales were recounted about the Bouto, as the large Dolphin of the Amazons is called. One of them was to the effect that a Bouto once had the habit of assuming the shape of a beautiful woman, with hair hanging loose to her heels, and walking ashore at night in the streets of Ega, to entice the young men down to the water. If any one was so much smitten as to follow her to the waterside, she grasped her victim round the waist and plunged beneath the waves with a triumphant cry. No animal in the Amazons region is the subject of so many fables as the Bouto; but it is probable these did not originate with the Indians, but with the Portuguese colonists. It was several years before I could induce a fisherman to harpoon Dolphins for me as specimens, for no one ever kills these animals voluntarily, although their fat is known to yield an excellent oil for lamps. The superstitious people believe that blindness would result from the use of this oil in lamps. I succeeded at length with Carepira, by offering him a high reward when his finances were at a very low point, but he repented of his deed ever afterwards, declaring that his luck had forsaken him from that day.

The next morning we again beat the pool. Although we had proof of there being a great number of turtles yet remaining, we had very poor success. The old Indians told us it would be so, for the turtles were 'ladino' (cunning), and would take no notice of the beating a second day. When the net was formed into a circle, and the men had jumped in, an alligator was found to be inclosed. No one was alarmed, the only fear expressed being that the imprisoned beast would tear the net. First one shouted, 'I have touched his head;' then another, 'he has scratched my leg;' one of the men, a lanky Miranha, was thrown off his balance, and then there was no end to the laughter and shouting. At last a youth of about fourteen years of age, on my calling to him from the bank to do so, seized the reptile by the tail, and held him tightly until, a little resistance being overcome, he was able to bring it ashore. The net was opened, and the boy slowly dragged the dangerous but cowardly beast to land through the muddy water, a distance of about a hundred yards. Meantime, I had cut a strong pole from a tree, and as soon as the alligator was drawn to solid ground, gave him a smart rap with it on the crown of his head, which killed him instantly. It was a good-sized individual, the jaws being considerably more than a foot long, and fully capable of snapping a man's leg in twain. The species was the large cayman, the Jacare-uassu of the Amazonian Indians (Jacare nigra).

On the third day, we sent our men in the boats to net turtles in a larger pool about five miles further down the river, and on the fourth, returned to Ega.

It will be well to mention here a few circumstances relative to the large Cayman, which, with the incident just narrated, afford illustrations of the cunning, cowardice, and ferocity of this reptile.

I have hitherto had but few occasions of mentioning alligators, although they exist by myriads in the waters of the Upper Amazons. Many different species are spoken of by the natives. I saw only three, and of these two only are common: one, the Jacare-tinga, a small kind (five feet long when full grown), having a long slender muzzle and a black-banded tail; the other, the Jacare-uassu, to which these remarks more especially relate and the third the Jacare-curua, mentioned in a former chapter. The Jacare-uassu, or large Cayman, grows to a length of eighteen or twenty feet, and attains an enormous bulk. Like the turtles, the alligator has its annual migrations, for it retreats to the interior pools and flooded forests in the wet season, and descends to the main river in the dry season. During the months of high water, therefore, scarcely a single individual is to be seen in the main river. In the middle part of the Lower Amazons, about Obydos and Villa Nova, where many of the lakes with their channels of communication with the trunk stream dry up in the fine months, the alligator buries itself in the mud and becomes dormant, sleeping till the rainy season returns. On the Upper Amazons, where the dry season is never excessive, it has not this habit, but is lively all the year round. It is scarcely exaggerating to say that the waters of the Solimoens are as well stocked with large alligators in the dry season, as a ditch in England is in summer with tadpoles. During a journey of five days which I once made in the Upper Amazons steamer, in November, alligators were seen along the coast almost every step of the way, and the passengers amused themselves, from morning till night, by firing at them with rifle and ball. They were very numerous in the still bays, where the huddled crowds jostled together, to the great rattling of their coats of mail, as the steamer passed.

The natives at once despise and fear the great cayman. I once spent a month at Caicara, a small village of semi-civilised Indians, about twenty miles to the west of Ega. My entertainer, the only white in the place, and one of my best and most constant friends, Senor Innocencio Alves Faria, one day proposed a half-day's fishing with net in the lake – the expanded bed of the small river on which the village is situated. We set out in an open boat with six Indians and two of Innocencio's children. The water had sunk so low that the net had to be taken out into the middle by the Indians, whence at the first draught, two medium-sized alligators were brought to land. They were disengaged from the net and allowed, with the coolest unconcern, to return to the water, although the two children were playing in it not many yards off. We continued fishing, Innocencio and I lending a helping hand, and each time drew a number of the reptiles of different ages and sizes, some of them Jacare-tingas; the lake, in fact, swarmed with alligators. After taking a very large quantity of fish, we prepared to return, and the Indians, at my suggestion, secured one of the alligators with the view of letting it loose amongst the swarms of dogs in the village. An individual was selected about eight feet long – one man holding his head and another his tail, whilst a third took a few lengths of a flexible liana, and deliberately bound the jaws and the legs. Thus secured, the beast was laid across the benches of the boat on which we sat during the hour and a half's journey to the settlement. We were rather crowded, but our amiable passenger gave us no trouble during the transit. On reaching the village, we took the animal into the middle of the green, in front of the church, where the dogs were congregated, and there gave him his liberty, two of us arming ourselves with long poles to intercept him if he should make for the water, and the others exciting the dogs. The alligator showed great terror, although the dogs could not be made to advance, and made off at the top of its speed for the water, waddling like a duck. We tried to keep him back with the poles, but he became enraged, and seizing the end of the one I held in his jaws, nearly wrenched it from my grasp. We were obliged, at length, to kill him to prevent his escape.

These little incidents show the timidity or cowardice of the alligator. He never attacks man when his intended victim is on his guard; but he is cunning enough to know when this may be done with impunity – of this we had proof at Caicara, a few days afterwards. The river had sunk to a very low point, so that the port and bathing-place of the village now lay at the foot of a long sloping bank, and a large cayman made his appearance in the shallow and muddy water. We were all obliged to be very careful in taking our bath; most of the people simply using a calabash, pouring the water over themselves while standing on the brink. A large trading canoe, belonging to a Barra merchant named Soares, arrived at this time, and the Indian crew, as usual, spent the first day or two after their coming into port in drunkenness and debauchery ashore. One of the men, during the greatest heat of the day, when almost everyone was enjoying his afternoon's nap, took it into his head while in a tipsy state to go down alone to bathe. He was seen only by the Juiz de Paz, a feeble old man who was lying in his hammock in the open verandah at the rear of his house on the top of the bank, and who shouted to the besotted Indian to beware of the alligator. Before he could repeat his warning, the man stumbled, and a pair of gaping jaws, appearing suddenly above the surface, seized him round the waist and drew him under the water. A cry of agony 'Ai Jesus!' was the last sign made by the wretched victim. The village was aroused: the young men with praiseworthy readiness seized their harpoons and hurried down to the bank; but, of course it was too late, a winding track of blood on the surface of the water was all that could be seen. They embarked, however, in montarias, determined upon vengeance; the monster was traced, and when, after a short lapse of time, he came up to breathe – one leg of the man sticking out from his jaws – was despatched with bitter curses.

The last of these minor excursions which I shall narrate, was made (again in company of Senor Cardozo, with the addition of his housekeeper Senora Felippa) in the season when all the population of the villages turns out to dig up turtle eggs, and revel on the praias. Placards were posted on the church doors at Ega, announcing that the excavation on Shimuni would commence on the 17th of October, and on Catua, sixty miles below Shimuni, on the 25th. We set out on the 16th, and passed on the road, in our well-manned igarite, a large number of people – men, women, and children in canoes of all sizes – wending their way as if to a great holiday gathering. By the morning of the 17th, some 400 persons were assembled on the borders of the sand-bank; each family having erected a rude temporary shed of poles and palm leaves to protect themselves from the sun and rain. Large copper kettles to prepare the oil, and hundreds of red earthenware jars, were scattered about on the sand.

The excavation of the taboleiro, collecting the eggs and purifying the oil, occupied four days. All was done on a system established by the old Portuguese governors, probably more than a century ago. The commandante first took down the names of all the masters of households, with the number of persons each intended to employ in digging; he then exacted a payment of 140 reis (about fourpence) a head, towards defraying the expense of sentinels. The whole were then allowed to go to the taboleiro. They arranged themselves around the circle, each person armed with a paddle to be used as a spade, and then all began simultaneously to dig on a signal being given – the roll of drums – by order of the commandante. It was an animating sight to behold the wide circle of rival diggers throwing up clouds of sand in their energetic labours, and working gradually towards the centre of the ring. A little rest was taken during the great heat of midday, and in the evening the eggs were carried to the huts in baskets. By the end of the second day, the taboleiro was exhausted; large mounds of eggs, some of them four to five feet in height, were then seen by the side of each hut, the produce of the labours of the family.

In the hurry of digging, some of the deeper nests are passed over; to find these out, the people go about provided with a long steel or wooden probe, the presence of the eggs being discoverable by the ease with which the spit enters the sand. When no more eggs are to be found, the mashing process begins. The egg, it may be mentioned, has a flexible or leathery shell; it is quite round, and somewhat larger than a hen's egg. The whole heap is thrown into an empty canoe and mashed with wooden prongs; but sometimes naked Indians and children jump into the mass and tread it down, besmearing themselves with yolk and making about as filthy a scene as can well be imagined. This being finished, water is poured into the canoe, and the fatty mess then left for a few hours to be heated by the sun, on which the oil separates and rises to the surface. The floating oil is afterwards skimmed off with long spoons, made by tying large mussel-shells to the end of rods, and purified over the fire in copper kettles.

The destruction of turtle eggs every year by these proceedings is enormous. At least 6000 jars, holding each three gallons of the oil, are exported annually from the Upper Amazons and the Madeira to Para, where it is used for lighting, frying fish, and other purposes. It may be fairly estimated that 2000 more jarsfull are consumed by the inhabitants of the villages on the river. Now, it takes at least twelve basketsful of eggs, or about 6000 by the wasteful process followed, to make one jar of oil. The total number of eggs annually destroyed amounts, therefore, to 48,000,000. As each turtle lays about 120, it follows that the yearly offspring of 400,000 turtles is thus annihilated. A vast number, nevertheless, remain undetected; and these would probably be sufficient to keep the turtle population of these rivers up to the mark, if the people did not follow the wasteful practice of lying in wait for the newly-hatched young, and collecting them by thousands for eating – their tender flesh and the remains of yolk in their entrails being considered a great delicacy. The chief natural enemies of the turtle are vultures and alligators, which devour the newly-hatched young as they descend in shoals to the water. These must have destroyed an immensely greater number before the European settlers began to appropriate the eggs than they do now. It is almost doubtful if this natural persecution did not act as effectively in checking the increase of the turtle as the artificial destruction now does. If we are to believe the tradition of the Indians, however, it had not this result; for they say that formerly the waters teemed as thickly with turtles as the air now does with mosquitoes. The universal opinion of the settlers on the Upper Amazons is, that the turtle has very greatly decreased in numbers, and is still annually decreasing.

We left Shimuni on the 20th with quite a flotilla of canoes, and descended the river to Catua, an eleven hours' journey by paddle and current. Catua is about six miles long, and almost entirely encircled by its praia. The turtles had selected for their egg-laying a part of the sand-bank which was elevated at least twenty feet above the present level of the river; the animals, to reach the place, must have crawled up a slope. As we approached the island, numbers of the animals were seen coming to the surface to breathe, in a small shoaly bay. Those who had light montarias sped forward with bows and arrows to shoot them. Carepira was foremost, having borrowed a small and very unsteady boat, of Cardozo, and embarked in it with his little son. After bagging a couple of turtles, and while hauling in a third, he overbalanced himself; the canoe went over, and he with his child had to swim for their lives in the midst of numerous alligators, about a mile from the land. The old man had to sustain a heavy fire of jokes from his companions for several days after this mishap. Such accidents are only laughed at by this almost amphibious people.

The number of persons congregated on Catua was much greater than on Shimuni, as the population of the banks of several neighbouring lakes were here added. The line of huts and sheds extended half a mile, and several large sailing vessels were anchored at the place. The commandante was Senor Macedo, the Indian blacksmith of Ega before mentioned, who maintained excellent order during the fourteen days the process of excavation and oil manufacture lasted. There were also many primitive Indians here from the neighbouring rivers, among them a family of Shumanas, good-tempered, harmless people from the Lower Japura. All of them were tattooed around the mouth, the bluish tint forming a border to the lips, and extending in a line on the cheeks towards the ear on each side. They were not quite so slender in figure as the Passes of Pedro-uassu's family; but their features deviated quite as much as those of the Passes from the ordinary Indian type. This was seen chiefly in the comparatively small mouth, pointed chin, thin lips, and narrow, high nose. One of the daughters, a young girl of about seventeen years of age, was a real beauty. The colour of her skin approached the light tanned shade of the Mameluco women; her figure was almost faultless, and the blue mouth, instead of being a disfigurement, gave quite a captivating finish to her appearance. Her neck, wrists, and ankles were adorned with strings of blue beads. She was, however, extremely bashful, never venturing to look strangers in the face, and never quitting, for many minutes together, the side of her father and mother. The family had been shamefully swindled by some rascally trader on another praia; and, on our arrival, came to lay their case before Senor Cardozo, as the delegado of police of the district. The mild way in which the old man, without a trace of anger, stated his complaint in imperfect Tupi quite enlisted our sympathies in his favour. But Cardozo could give him no redress; he invited the family, however, to make their rancho near to ours, and in the end gave them the highest price for the surplus oil which they manufactured.

It was not all work at Catua; indeed there was rather more play than work going on. The people make a kind of holiday of these occasions. Every fine night parties of the younger people assembled on the sands, and dancing and games were carried on for hours together. But the requisite liveliness for these sports was never got up without a good deal of preliminary rum-drinking. The girls were so coy that the young men could not get sufficient partners for the dances without first subscribing for a few flagons of the needful cashaca. The coldness of the shy Indian and Mameluco maidens never failed to give way after a little of this strong drink, but it was astonishing what an immense deal they could take of it in the course of an evening. Coyness is not always a sign of innocence in these people, for most of the half-caste women on the Upper Amazons lead a little career of looseness before they marry and settle down for life; and it is rather remarkable that the men do not seem to object much to their brides having had a child or two by various fathers before marriage. The women do not lose reputation unless they become utterly depraved, but in that case they are condemned pretty strongly by public opinion. Depravity is, however, rare, for all require more or less to be wooed before they are won. I did not see (although I mixed pretty freely with the young people) any breach of propriety on the praias. The merry-makings were carried on near the ranchos, where the more staid citizens of Ega, husbands with their wives and young daughters, all smoking gravely out of long pipes, sat in their hammocks and enjoyed the fun. Towards midnight we often heard, in the intervals between jokes and laughter, the hoarse roar of jaguars prowling about the jungle in the middle of the praia. There were several guitar-players among the young men, and one most persevering fiddler – so there was no lack of music.

The favourite sport was the Pira-purasseya, or fish-dance, one of the original games of the Indians, though now probably a little modified. The young men and women, mingling together, formed a ring, leaving one of their number in the middle, who represented the fish. They then all marched round, Indian file, the musicians mixed up with the rest, singing a monotonous but rather pretty chorus, the words of which were invented (under a certain form) by one of the party who acted as leader. This finished, all joined hands, and questions were put to the one in the middle, asking what kind of fish he or she might be. To these the individual has to reply. The end of it all is that he makes a rush at the ring, and if he succeeds in escaping, the person who allowed him to do so has to take his place; the march and chorus then recommences, and so the game goes on hour after hour. Tupi was the language mostly used, but sometimes Portuguese was sung and spoken. The details of the dance were often varied. Instead of the names of fishes being called over by the person in the middle, the name of some animal, flower, or other object was given to every fresh occupier of the place. There was then good scope for wit in the invention of nicknames, and peals of laughter would often salute some particularly good hit. Thus a very lanky young man was called the Magoary, or the grey stork; a moist grey-eyed man with a profile comically suggestive of a fish was christened Jaraki (a kind of fish), which was considered quite a witty sally; a little Mameluco girl, with light-coloured eyes and brown hair, got the gallant name of Rosa Blanca, or the white rose; a young fellow who had recently singed his eye brows by the explosion of fireworks, was dubbed Pedro queimado (burnt Peter); in short every one got a nick-name, and each time the cognomen was introduced into the chorus as the circle marched round.

Our rancho was a large one, and was erected in a line with the others near the edge of the sand-bank which sloped rather abruptly to the water. During the first week the people were all, more or less, troubled by alligators. Some half-dozen full-grown ones were in attendance off the praia, floating about on the lazily-flowing, muddy water. The dryness of the weather had increased since we had left Shimuni, the currents had slackened, and the heat in the middle part of the day was almost insupportable. But no one could descend to bathe without being advanced upon by one or other of these hungry monsters. There was much offal cast into the river, and this, of course, attracted them to the place. One day I amused myself by taking a basketful of fragments of meat beyond the line of ranchos, and drawing the alligators towards me by feeding them. They behaved pretty much as dogs do when fed – catching the bones I threw them in their huge jaws, and coming nearer and showing increased eagerness after every morsel. The enormous gape of their mouths, with their blood-red lining and long fringes of teeth, and the uncouth shapes of their bodies, made a picture of unsurpassable ugliness. I once or twice fired a heavy charge of shot at them, aiming at the vulnerable part of their bodies, which is a small space situated behind the eyes, but this had no other effect than to make them give a hoarse grunt and shake themselves; they immediately afterwards turned to receive another bone which I threw to them.

Every day these visitors became bolder; at length they reached a pitch of impudence that was quite intolerable. Cardozo had a poodle dog named Carlito, which some grateful traveller whom he had befriended had sent him from Rio Janeiro. He took great pride in this dog, keeping it well sheared, and preserving his coat as white as soap and water could make it. We slept in our rancho in hammocks slung between the outer posts; a large wood fire (fed with a kind of wood abundant on the banks of the river, which keeps alight all night) being made in the middle, by the side of which slept Carlito on a little mat. Well, one night I was awoke by a great uproar. It was caused by Cardozo hurling burning firewood with loud curses at a huge cayman which had crawled up the bank and passed beneath my hammock (being nearest the water) towards the place where Carlito lay. The dog had raised the alarm in time; the reptile backed out and tumbled down the bank to the water, the sparks from the brands hurled at him flying from his bony hide. To our great surprise the animal (we supposed it to be the same individual) repeated his visit the very next night, this time passing round to the other side of our shed. Cardozo was awake, and threw a harpoon at him, but without doing him any harm. After this it was thought necessary to make an effort to check the alligators; a number of men were therefore persuaded to sally forth in their montarias and devote a day to killing them.

The young men made several hunting excursions during the fourteen days of our stay on Catua, and I, being associated with them in all their pleasures, made generally one of the party. These were, besides, the sole occasions on which I could add to my collections, while on these barren sands. Only two of these trips afforded incidents worth relating.

The first, which was made to the interior of the wooded island of Catua, was not a very successful one. We were twelve in number, all armed with guns and long hunting-knives. Long before sunrise, my friends woke me up from my hammock, where I lay, as usual, in the clothes worn during the day; and after taking each a cup-full of cashaca and ginger (a very general practice in early morning on the sand-banks), we commenced our walk. The waning moon still lingered in the clear sky, and a profound stillness pervaded the sleeping camp, forest, and stream. Along the line of ranchos glimmered the fires made by each party to dry turtle-eggs for food, the eggs being spread on little wooden stages over the smoke. The distance to the forest from our place of starting was about two miles, being nearly the whole length of the sand-bank, which was also a very broad one – the highest part, where it was covered with a thicket of dwarf willows, mimosas, and arrow grass, lying near the ranchos. We loitered much on the way, and the day dawned whilst we were yet on the road, the sand at this early hour feeling quite cold to the naked feet. As soon as we were able to distinguish things, the surface of the praia was seen to be dotted with small black objects. These were newly-hatched Aiyussa turtles, which were making their way in an undeviating line to the water, at least a mile distant. The young animal of this species is distinguishable from that of the large turtle and the Tracaja, by the edges of the breast-plate being raised on each side, so that in crawling it scores two parallel lines on the sand. The mouths of these little creatures were full of sand, a circumstance arising from their having to bite their way through many inches of superincumbent sand to reach the surface on emerging from the buried eggs. It was amusing to observe how constantly they turned again in the direction of the distant river, after being handled and set down on the sand with their heads facing the opposite quarter. We saw also several skeletons of the large cayman (some with the horny and bony hide of the animal nearly perfect) embedded in the sand; they reminded me of the remains of Ichthyosauri fossilised in beds of lias, with the difference of being buried in fine sand instead of in blue mud. I marked the place of one which had a well-preserved skull, and the next day returned to secure it. The specimen is now in the British Museum collection. There were also many footmarks of jaguars on the sand.

We entered the forest, as the sun peeped over the tree-tops far away down river. The party soon after divided, I keeping with a section which was led by Bento, the Ega carpenter, a capital woodsman. After a short walk we struck the banks of a beautiful little lake, having grassy margins and clear dark water, on the surface of which floated thick beds of water-lilies. We then crossed a muddy creek or watercourse that entered the lake, and then found ourselves on a restinga, or tongue of land between two waters. By keeping in sight of one or the other of these, there was no danger of our losing our way – all other precautions were therefore unnecessary. The forest was tolerably clear of underwood, and consequently, easy to walk through. We had not gone far before a soft, long-drawn whistle was heard aloft in the trees, betraying the presence of Mutums (Curassow birds). The crowns of the trees, a hundred feet or more over our heads, were so closely interwoven that it was difficult to distinguish the birds – the practised eye of Bento, however, made them out, and a fine male was shot from the flock, the rest flying away and alighting at no great distance. The species was the one of which the male has a round red ball on its beak (Crax globicera). The pursuit of the others led us a great distance, straight towards the interior of the island, in which direction we marched for three hours, having the lake always on our right.

Arriving at length at the head of the lake, Bento

struck off to the left across the restinga, and we then soon came upon a treeless space choked up with tall grass, which appeared to be the dried-up bed of another lake. Our leader was obliged to climb a tree to ascertain our position, and found that the clear space was part of the creek, whose mouth we had crossed lower down. The banks were clothed with low trees, nearly all of one species, a kind of araca (Psidium), and the ground was carpeted with a slender delicate grass, now in flower. A great number of crimson and vermilion-coloured butterflies (Catagramma Peristera, male and female) were settled on the smooth, white trunks of these trees. I had also here the great pleasure of seeing for the first time, the rare and curious Umbrella Bird (Cephalopterus ornatus), a species which resembles in size, colour, and appearance our common crow, but is decorated with a crest of long, curved, hairy feathers having long bare quills, which, when raised, spread themselves out in the form of a fringed sunshade over the head. A strange ornament, like a pelerine, is also suspended from the neck, formed by a thick pad of glossy steel-blue feathers, which grow on a long fleshy lobe or excrescence. This lobe is connected (as I found on skinning specimens) with an unusual development of the trachea and vocal organs, to which the bird doubtless owes its singularly deep, loud, and long-sustained fluty note. The Indian name of this strange creature is Uira-mimbeu, or fife-bird, (Mimbeu is the Indian name for a rude kind of pan-pipes used by the Caishanas and other tribes) in allusion to the tone of its voice. We had the good luck, after remaining quiet a short time, to hear its performance. It drew itself up on its perch, spread widely the umbrella-formed crest, dilated and waved its glossy breast-lappet, and then, in giving vent to its loud piping note, bowed its head slowly forwards. We obtained a pair, male and female; the female has only the rudiments of the crest and lappet, and is duller-coloured altogether than the male. The range of this bird appears to be quite confined to the plains of the Upper Amazons (especially the Ygapo forests), not having been found to the east of the Rio Negro.

Bento and our other friends being disappointed in finding no more Curassows, or indeed any other species of game, now resolved to turn back. On reaching the edge of the forest, we sat down and ate our dinners under the shade – each man having brought a little bag containing a few handsfull of farinha, and a piece of fried fish or roast turtle. We expected our companions of the other division to join us at midday, but after waiting till past one o'clock without seeing anything of them (in fact, they had returned to the huts an hour or two previously), we struck off across the praia towards the encampment. An obstacle here presented itself on which we had not counted. The sun had shone all day through a cloudless sky untempered by a breath of wind, and the sands had become heated by it to a degree that rendered walking over them with our bare feet impossible. The most hardened footsoles of the party could not endure the burning soil. We made several attempts; we tried running, having wrapped the cool leaves of Heliconiae round our feet, but in no way could we step forward many yards. There was no means of getting back to our friends before night, except going round the praia, a circuit of about four miles, and walking through the water or on the moist sand. To get to the waterside from the place where we then stood was not difficult, as a thick bed of a flowering shrub, called tintarana, an infusion of the leaves of which is used to dye black, lay on that side of the sand-bank. Footsore and wearied, burthened with our guns, and walking for miles through the tepid shallow water under the brain-scorching vertical sun, we had, as may be imagined, anything but a pleasant time of it. I did not, however, feel any inconvenience afterwards. Everyone enjoys the most lusty health while living this free and wild life on the rivers.

The other hunting trip which I have alluded to was undertaken in company with three friendly young half-castes. Two of them were brothers, namely, Joao (John) and Zephyrino Jabuti: Jabuti, or tortoise, being a nickname which their father had earned for his slow gait, and which, as is usual in this country, had descended as the surname of the family. The other was Jose Frazao, a nephew of Senor Chrysostomo, of Ega, an active, clever, and manly young fellow, whom I much esteemed. He was almost a white – his father being a Portuguese and his mother a Mameluca. We were accompanied by an Indian named Lino, and a Mulatto boy, whose office was to carry our game.

Our proposed hunting-ground on this occasion lay across the water, about fifteen miles distant. We set out in a small montaria, at four o'clock in the morning, again leaving the encampment asleep, and travelled at a good pace up the northern channel of the Solimoens, or that lying between the island Catua and the left bank of the river. The northern shore of the island had a broad sandy beach reaching to its western extremity. We gained our destination a little after daybreak; this was the banks of the Carapanatuba, (meaning, in Tupi, the river of many mosquitoes: from carapana, mosquito, and ituba, many.) a channel some 150 yards in width, which, like the Anana already mentioned, communicates with the Cupiyo. To reach this we had to cross the river, here nearly two miles wide. Just as day dawned we saw a Cayman seize a large fish, a Tambaki, near the surface; the reptile seemed to have a difficulty in securing its prey, for it reared itself above the water, tossing the fish in its jaws and making a tremendous commotion. I was much struck also by the singular appearance presented by certain diving birds having very long and snaky necks (the Plotus Anhinga). Occasionally a long serpentine form would suddenly wriggle itself to a height of a foot and a half above the glassy surface of the water, producing such a deceptive imitation of a snake that at first I had some difficulty in believing it to be the neck of a bird; it did not remain long in view, but soon plunged again beneath the stream.

We ran ashore in a most lonely and gloomy place, on a low sand-bank covered with bushes, secured the montaria to a tree, and then, after making a very sparing breakfast of fried fish and mandioca meal, rolled up our trousers and plunged into the thick forest, which here, as everywhere else, rose like a lofty wall of foliage from the narrow strip of beach. We made straight for the heart of the land, John Jabuti leading, and breaking off at every few steps a branch of the lower trees, so that we might recognise the path on our return. The district was quite new to all my companions, and being on a coast almost totally uninhabited by human beings for some 300 miles, to lose our way would have been to perish helplessly. I did not think at the time of the risk we ran of having our canoe stolen by passing Indians, unguarded montarias being never safe even in the ports of the villages, Indians apparently considering them common property, and stealing them without any compunction. No misgivings clouded the lightness of heart with which we trod forward in warm anticipation of a good day's sport.

The tract of forest through which we passed was Ygapo, but the higher parts of the land formed areas which went only a very few inches under water in the flood season. It consisted of a most bewildering diversity of grand and beautiful trees, draped, festooned, corded, matted, and ribboned with climbing plants, woody and succulent, in endless variety. The most prevalent palm was the tall Astryocaryum Jauari, whose fallen spines made it necessary to pick our way carefully over the ground, as we were all barefooted. There was not much green underwood, except in places where Bamboos grew; these formed impenetrable thickets of plumy foliage and thorny, jointed stems, which always compelled us to make a circuit to avoid them. The earth elsewhere was encumbered with rotting fruits, gigantic bean-pods, leaves, limbs, and trunks of trees; fixing the impression of its being the cemetery as well as the birthplace of the great world of vegetation overhead. Some of the trees were of prodigious height. We passed many specimens of the Moratinga, whose cylindrical trunks, I dare not say how many feet in circumference, towered up and were lost amidst the crowns of the lower trees, their lower branches, in some cases, being hidden from our view. Another very large and remarkable tree was the Assacu (Sapium aucuparium). A traveller on the Amazons, mingling with the people, is sure to hear much of the poisonous qualities of the juices of this tree. Its bark exudes, when hacked with a knife, a milky sap, which is not only a fatal poison when taken internally, but is said to cause incurable sores if simply sprinkled on the skin. My companions always gave the Assacu a wide berth when we passed one. The tree looks ugly enough to merit a bad name, for the bark is of a dingy olive colour, and is studded with short and sharp, venomous-looking spines.

After walking about half a mile we came upon a dry watercourse, where we observed, first, the old foot-marks of a tapir, and, soon after, on the margin of a curious circular hole full of muddy water, the fresh tracks of a Jaguar. This latter discovery was hardly made when a rush was heard amidst the bushes on the top of a sloping bank on the opposite side of the dried creek. We bounded forward; it was, however, too late, for the animal had sped in a few minutes far out of our reach. It was clear we had disturbed, on our approach, the Jaguar, while quenching his thirst at the water-hole. A few steps further on we saw the mangled remains of an alligator (the Jacaretinga). The head, forequarters, and bony shell were the only parts which remained; but the meat was quite fresh, and there were many footmarks of the Jaguar around the carcase – so that there was no doubt this had formed the solid part of the animal's breakfast. My companions now began to search for the alligator's nest, the presence of the reptile so far from the river being accountable for on no other ground than its maternal solicitude for its eggs. We found, in fact, the nest at the distance of a few yards from the place. It was a conical pile of dead leaves, in the middle of which twenty eggs were buried. These were of elliptical shape, considerably larger than those of a duck, and having a hard shell of the texture of porcelain, but very rough on the outside. They make a loud sound when rubbed together, and it is said that it is easy to find a mother alligator in the Ygapo forests by rubbing together two eggs in this way, she being never far off, and attracted by the sounds.

I put half-a-dozen of the alligator's eggs in my game-bag for specimens, and we then continued on our way. Lino, who was now first, presently made a start backwards, calling out 'Jararaca!' This is the name of a poisonous snake (genus Craspedocephalus), which is far more dreaded by the natives than Jaguar or Alligator. The individual seen by Lino lay coiled up at the foot of a tree, and was scarcely distinguishable, on account of the colours of its body being assimilated to those of the fallen leaves. Its hideous, flat triangular head, connected with the body by a thin neck, was reared and turned towards us: Frazao killed it with a charge of shot, shattering it completely, and destroying, to my regret, its value as a specimen. In conversing on the subject of Jararacas as we walked onwards, every one of the party was ready to swear that this snake attacks man without provocation, leaping towards him from a considerable distance when he approaches. I met, in the course of my daily rambles in the woods, many Jararacas, and once or twice very narrowly escaped treading on them, but never saw them attempt to spring. On some subjects the testimony of the natives of a wild country is utterly worthless. The bite of the Jararacas is generally fatal. I knew of four or five instances of death from it, and only of one clear case of recovery after being bitten; but in that case the person was lamed for life.

We walked over moderately elevated and dry ground for about a mile, and then descended (three or four feet only) to the dry bed of another creek. This was pierced in the same way as the former water-course, with round holes full of muddy water. They occurred at intervals of a few yards, and had the appearance of having been made by the hand of man. The smallest were about two feet, the largest seven or eight feet in diameter. As we approached the most extensive of the larger ones, I was startled at seeing a number of large serpent-like heads bobbing about the surface. They proved to be those of electric eels, and it now occurred to me that the round holes were made by these animals working constantly round and round in the moist, muddy soil. Their depth (some of them were at least eight feet deep) was doubtless due also to the movements of the eels in the soft soil, and accounted for their not drying up, in the fine season, with the rest of the creek. Thus, while alligators and turtles in this great inundated forest region retire to the larger pools during the dry season, the electric eels make for themselves little ponds in which to pass the season of drought.

My companions now cut each a stout pole, and proceeded to eject the eels in order to get at the other fishes, with which they had discovered the ponds to abound. I amused them all very much by showing how the electric shock from the eels could pass from one person to another. We joined hands in a line while I touched the biggest and freshest of the animals on the head with the point of my hunting-knife. We found that this experiment did not succeed more than three times with the same eel when out of the water; for, the fourth time the shock was scarcely perceptible. All the fishes found in the holes (besides the eels) belonged to one species, a small kind of Acari, or Loricaria, a group whose members have a complete bony integument. Lino and the boy strung them together through the gills with slender sipos, and hung them on the trees to await our return later in the day.

Leaving the bed of the creek, we marched onwards, always towards the centre of the land, guided by the sun, which now glimmered through the thick foliage overhead. About eleven o'clock we saw a break in the forest before us, and presently emerged on the banks of a rather large sheet of water. This was one of the interior pools of which there are so many in this district. The margins were elevated some few feet, and sloped down to the water, the ground being hard and dry to the water's edge, and covered with shrubby vegetation. We passed completely round this pool, finding the crowns of the trees on its borders tenanted by curassow birds, whose presence was betrayed as usual by the peculiar note which they emit. My companions shot two of them. At the further end of the lake lay a deep watercourse, which we traced for about half a mile, and found to communicate with another and smaller pool. This second one evidently swarmed with turtles, as we saw the snouts of many peering above the surface of the water: the same had not been seen in the larger lake, probably because we had made too much noise in hailing our discovery on approaching its banks. My friends made an arrangement on the spot for returning to this pool, after the termination of the egg harvest on Catua.

In recrossing the space between the two pools, we heard the crash of monkeys in the crowns of trees overhead. The chase of these occupied us a considerable time. Jose fired at length at one of the laggards of the troop, and wounded him. He climbed pretty nimbly towards a denser part of the tree, and a second and third discharge failed to bring him down. The poor maimed creature then trailed his limbs to one of the topmost branches, where we descried him soon after, seated and picking the entrails from a wound in his abdomen – a most heart-rending sight. The height from the ground to the bough on which he was perched could not have been less than 150 feet, and we could get a glimpse of him only by standing directly underneath, and straining our eyes upwards. We killed him at last by loading our best gun with a careful charge, and resting the barrel against the treetrunk to steady the aim. A few shots entered his chin, and he then fell heels over head screaming to the ground. Although it was I who gave the final shot, this animal did not fall to my lot in dividing the spoils at the end of the day. I regret now not having preserved the skin, as it belonged to a very large species of Cebus, and one which I never met with afterwards.

It was about one o'clock in the afternoon when we again reached the spot where we had first struck the banks of the larger pool. We hitherto had but poor sport, so after dining on the remains of our fried fish and farinha, and smoking our cigarettes, the apparatus for making which, including bamboo tinder-box and steel and flint for striking a light, being carried by every one always on these expeditions, we made off in another (westerly) direction through the forest to try to find better hunting-ground. We quenched our thirst with water from the pool, which I was rather surprised to find quite pure. These pools are, of course, sometimes fouled for a time by the movements of alligators and other tenants in the fine mud which settles at the bottom, but I never observed a scum of confervae or traces of oil revealing animal decomposition on the surface of these waters, nor was there ever any foul smell perceptible. The whole of this level land, instead of being covered with unwholesome swamps emitting malaria, forms in the dry season (and in the wet also) a most healthy country. How elaborate must be the natural processes of self-purification in these teeming waters!

On our fresh route we were obliged to cut our way through a long belt of bamboo underwood, and not being so careful of my steps as my companions, I trod repeatedly on the flinty thorns which had fallen from the bushes, finishing by becoming completely lame, one thorn having entered deeply the sole of my foot. I was obliged to be left behind – Lino, the Indian, remaining with me. The careful fellow cleaned my wounds with his saliva, placed pieces of isca (the felt-like substance manufactured by ants) on them to staunch the blood, and bound my feet with tough bast to serve as shoes, which he cut from the bark of a Monguba tree. He went about his work in a very gentle way and with much skill, but was so sparing of speech that I could scarcely get answers to the questions I put to him. When he had done I was able to limp about pretty nimbly. An Indian when he performs a service of this kind never thinks of a reward. I did not find so much disinterestedness in negro slaves or half-castes. We had to wait two hours for the return of our companions; during part of this time I was left quite alone, Lino having started off into the jungle after a peccary (a kind of wild hog) which had come near to where we sat, but on seeing us had given a grunt and bounded off into the thickets. At length our friends hove in sight, loaded with game; having shot twelve curassows and two cujubims (Penelope Pipile), a handsome black fowl with a white head, which is arboreal in its habits like the rest of this group of Gallinaceous birds inhabiting the South American forests. They had discovered a third pool containing plenty of turtles. Lino rejoined us at the same time, having missed the peccary, but in compensation shot a Quandu, or porcupine. The mulatto boy had caught alive in the pool a most charming little water-fowl, a species of grebe. It was somewhat smaller than a pigeon, and had a pointed beak; its feet were furnished with many intricate folds or frills of skin instead of webs, and resembled very much those of the gecko lizards. The bird was kept as a pet in Jabuti's house at Ega for a long time afterwards, where it became accustomed to swim about in a common hand-basin full of water, and was a great favourite with everybody.

We now retraced our steps towards the water-side, a weary walk of five or six miles, reaching our canoe by half-past five o'clock, or a little before sunset. It was considered by everyone at Catua that we had had an unusually good day's sport. I never knew any small party to take so much game in one day in these forests, over which animals are everywhere so widely and sparingly scattered. My companions were greatly elated, and on approaching the encampment at Catua, made a great commotion with their paddles to announce their successful return, singing in their loudest key one of the wild choruses of the Amazonian boatmen.

The excavation of eggs and preparation of the oil being finished, we left Catua on the 3rd of November. Carepira, who was now attached to Cardozo's party, had discovered another lake rich in turtles, about twelve miles distant, in one of his fishing rambles, and my friend resolved, before returning to Ega, to go there with his nets and drag it as we had formerly done in the Aningal. Several Mameluco families of Ega begged to accompany us to share the labours and booty; the Shumana family also joined the party; we therefore, formed a large body, numbering in all eight canoes and fifty persons.

The summer season was now breaking up; the river was rising; the sky was almost constantly clouded, and we had frequent rains. The mosquitoes also, which we had not felt while encamped on the sand-banks, now became troublesome. We paddled up the north-westerly channel, and arrived at a point near the upper end of Catua at ten o'clock p.m. There was here a very broad beach of untrodden white sand, which extended quite into the forest, where it formed rounded hills and hollows like sand dunes, covered with a peculiar vegetation: harsh, reedy grasses, and low trees matted together with lianas, and varied with dwarf spiny palms of the genus Bactris. We encamped for the night on the sands, finding the place luckily free from mosquitoes. The different portions of the party made arched coverings with the toldos or maranta-leaf awnings of their canoes to sleep under, fixing the edges in the sand. No one, however, seemed inclined to go to sleep, so after supper we all sat or lay around the large fires and amused ourselves. We had the fiddler with us; and in the intervals between the wretched tunes which he played, the usual amusement of story-telling beguiled the time: tales of hair-breadth escapes from jaguar, alligator, and so forth. There were amongst us a father and son who had been the actors, the previous year, in an alligator adventure on the edge of the praia we had just left. The son, while bathing, was seized by the thigh and carried under water – a cry was raised, and the father, rushing down the bank, plunged after the rapacious beast, which was diving away with his victim. It seems almost incredible that a man could overtake and master the large cayman in his own element; but such was the case in this instance, for the animal was reached and forced to release his booty by the man's thrusting his thumb into his eye. The lad showed us the marks of the alligator's teeth on his thigh. We sat up until past midnight listening to these stories and assisting the flow of talk by frequent potations of burnt rum. A large, shallow dish was filled with the liquor and fired; when it had burned for a few minutes, the flame was extinguished and each one helped himself by dipping a teacup into the vessel.

One by one the people dropped asleep, and then the quiet murmur of talk of the few who remained awake was interrupted by the roar of jaguars in the jungle about a furlong distant. There was not one only, but several of the animals. The older men showed considerable alarm and proceeded to light fresh fires around the outside of our encampment. I had read in books of travel of tigers coming to warm themselves by the fires of a bivouac, and thought my strong wish to witness the same sight would have been gratified tonight. I had not, however, such good fortune, although I was the last to go to sleep, and my bed was the bare sand under a little arched covering open at both ends. The jaguars, nevertheless, must have come very near during the night, for their fresh footmarks were numerous within a score yards of the place where we slept. In the morning I had a ramble along the borders of the jungle, and found the tracks very numerous and close together on the sandy soil.

We remained in this neighbourhood four days, and succeeded in obtaining many hundred turtles, but we were obliged to sleep two nights within the Carapanatuba channel. The first night passed rather pleasantly, for the weather was fine, and we encamped in the forest, making large fires and slinging our hammocks between the trees. The second was one of the most miserable nights I ever spent. The air was close, and a drizzling rain began to fall about midnight, lasting until morning. We tried at first to brave it out under the trees. Several very large fires were made, lighting up with ruddy gleams the magnificent foliage in the black shades around our encampment. The heat and smoke had the desired effect of keeping off pretty well the mosquitoes, but the rain continued until at length everything was soaked, and we had no help for it but to bundle off to the canoes with drenched hammocks and garments. There was not nearly room enough in the flotilla to accommodate so large a number of persons lying at full length; moreover the night was pitch dark, and it was quite impossible in the gloom and confusion to get at a change of clothing. So there we lay, huddled together in the best way we could arrange ourselves, exhausted with fatigue and irritated beyond all conception by clouds of mosquitoes. I slept on a bench with a sail over me, my wet clothes clinging to my body, and to increase my discomfort, close beside me lay an Indian girl, one of Cardozo's domestics, who had a skin disfigured with black diseased patches, and whose thick clothing, not having been washed during the whole time we had been out (eighteen days), gave forth a most vile effluvium.

We spent the night of the 7th of November pleasantly on the smooth sands, where the jaguars again serenaded us, and on the succeeding morning we commenced our return voyage to Ega. We first doubled the upper end of the island of Catua, and then struck off for the right bank of the Solimoens. The river was here of immense width, and the current was so strong in the middle that it required the most strenuous exertions on the part of our paddlers to prevent us from being carried miles away down the stream. At night we reached the Juteca, a small river which enters the Solimoens by a channel so narrow that a man might almost jump across it, but a furlong inwards expands into a very pretty lake several miles in circumference. We slept again in the forest, and again were annoyed by rain and mosquitoes; but this time Cardozo and I preferred remaining where we were to mingling with the reeking crowd in the boats. When the grey dawn arose a steady rain was still falling, and the whole sky had a settled, leaden appearance, but it was delightfully cool. We took our net into the lake and gleaned a good supply of delicious fish for breakfast. I saw at the upper end of this lake the native rice of this country growing wild.

The weather cleared up at ten o'clock a.m. At three p.m. we arrived at the mouth of the Cayambe, another tributary stream much larger than the Juteca. The channel of exit to the Solimoens was here also very narrow, but the expanded river inside is of vast dimensions: it forms a lake (I may safely venturc to say), several score miles in circumference. Although prepared for these surprises, I was quite taken aback in this case. We had been paddling all day along a monotonous shore, with the dreary Solimoens before us, here three to four miles broad, heavily rolling onward its muddy waters. We come to a little gap in the earthy banks, and find a dark, narrow inlet with a wall of forest overshadowing it on each side; we enter it, and at a distance of two or three hundred yards a glorious sheet of water bursts upon the view. The scenery of Cayambe is very picturesque. The land, on the two sides visible of the lake, is high, and clothed with sombre woods, varied here and there with a white-washed house, in the middle of a green patch of clearing, belonging to settlers. In striking contrast to these dark, rolling forests, is the vivid, light green and cheerful foliage of the woods on the numerous islets which rest like water-gardens on the surface of the lake. Flocks of ducks, storks, and snow-white herons inhabit these islets, and a noise of parrots with the tingling chorus of Tamburi-paras was heard from them as we passed. This has a cheering effect after the depressing stillness and absence of life in the woods on the margins of the main river.

Cardozo and I took a small boat and crossed the lake to visit one of the settlers, and on our return to our canoe, while in the middle of the lake, a squall suddenly arose in the direction towards which we were going, so that for a whole hour we were in great danger of being swamped. The wind blew away the awning and mats, and lashed the waters into foam, the waves rising to a great height. Our boat, fortunately, was excellently constructed, rising well towards the prow, so that with good steering we managed to head the billows as they arose, and escaped without shipping much water. We reached our igarite at sunset, and then made all speed to Curubaru, fifteen miles distant, to encamp for the night on the sands. We reached the praia at ten o'clock. The waters were now mounting fast upon the sloping beach, and we found on dragging the net next morning that fish was beginning to be scarce. Cardozo and his friends talked quite gloomily at breakfast time over the departure of the joyous verao [summer], and the setting in of the dull, hungry winter season.

At nine o'clock in the morning of the 10th of November a light wind from down river sprang up, and all who had sails hoisted them. It was the first time during our trip that we had had occasion to use our sails, so continual is the calm on this upper river. We bowled along merrily, and soon entered the broad channel lying between Baria and the mainland on the south bank. The wind carried us right into the mouth of the Teffe and at four o'clock p.m. we cast anchor in the port of Ega.

Toucans, Vampire Bats, Foraging Ants and Other Creatures

The Jupura. – A curious animal, known to naturalists as the Kinkajou, but called Jupura by the Indians of the Amazons, and considered by them as a kind of monkey, may be mentioned in this place. It is the Cercoleptes caudivolvus of zoologists, and has been considered by some authors as an intermediate form between the Lemur family of apes and the plantigrade Carnivora, or Bear family. It has decidedly no close relationship to either of the groups of American monkeys, having six cutting teeth to each jaw, and long claws instead of nails, with extremities of the usual shape of paws instead of hands. Its muzzle is conical and pointed, like that of many Lemurs of Madagascar; the expression of its countenance, and its habits and actions, are also very similar to those of Lemurs. Its tail is very flexible towards the tip, and is used to twine round branches in climbing. I did not see or hear anything of this animal while residing on the Lower Amazons, but on the banks of the Upper river, from the Teffe to Peru, it appeared to be rather common. It is nocturnal in its habits, like the owl-faced monkeys, although, unlike them, it has a bright, dark eye. I once saw it in considerable numbers, when on an excursion with an Indian companion along the low Ygapo shores of the Teffe, about twenty miles above Ega. We slept one night at the house of a native family living in the thick of the forest where a festival was going on and, there being no room to hang our hammocks under shelter on account of the number of visitors, we lay down on a mat in the open air, near a shed which stood in the midst of a grove of fruit-trees and pupunha palms. Past midnight, when all became still, after the uproar of holidaymaking, as I was listening to the dull, fanning sound made by the wings of impish hosts of vampire bats crowding round the Caju trees, a rustle commenced from the side of the woods, and a troop of slender, long-tailed animals were seen against the clear moonlit sky, taking flying leaps from branch to branch through the grove. Many of them stopped at the pupunha trees, and the hustling, twittering, and screaming, with sounds of falling fruits, showed how they were employed. I thought, at first, they were Nyctipitheci, but they proved to be Jupuras, for the owner of the house early next morning caught a young one, and gave it to me. I kept this as a pet animal for several weeks, feeding it on bananas and mandiocameal mixed with treacle. It became tame in a very short time, allowing itself to be caressed, but making a distinction in the degree of confidence it showed between myself and strangers. My pet was unfortunately killed by a neighbour's dog, which entered the room where it was kept. The animal is so difficult to obtain alive, its place of retreat in the daytime not being known to the natives, that I was unable to procure a second living specimen.

Bats. – The only other mammals that I shall mention

are the bats, which exist in very considerable numbers and variety in the forest, as well as in the buildings of the villages. Many small and curious species, living in the woods, conceal themselves by day under the broad leaf-blades of Heliconiae and other plants which grow in shady places; others cling to the trunks of trees. While walking through the forest in the daytime, especially along gloomy ravines, one is almost sure to startle bats from their sleeping-places; and at night they are often seen in great numbers flitting about the trees on the shady margins of narrow channels. I captured altogether, without giving especial attention to bats, sixteen different species at Ega.

The Vampire Bat. – The little grey blood-sucking Phyllostoma, mentioned in a former chapter as found in my chamber at Caripi, was not uncommon at Ega, where everyone believes it to visit sleepers and bleed them in the night. But the vampire was here by far the most abundant of the family of leaf-nosed bats. It is the largest of all the South American species, measuring twenty-eight inches in expanse of wing. Nothing in animal physiognomy can be more hideous than the countenance of this creature when viewed from the front; the large, leathery ears standing out from the sides and top of the head, the erect spear-shaped appendage on the tip of the nose, the grin and the glistening black eye, all combining to make up a figure that reminds one of some mocking imp of fable. No wonder that imaginative people have inferred diabolical instincts on the part of so ugly an animal. The vampire, however, is the most harmless of all bats, and its inoffensive character is well known to residents on the banks of the Amazons. I found two distinct species of it, one having the fur of a blackish colour, the other of a ruddy hue, and ascertained that both feed chiefly on fruits. The church at Ega was the headquarters of both kinds, I used to see them, as I sat at my door during the short evening twilights, trooping forth by scores from a large open window at the back of the altar, twittering cheerfully as they sped off to the borders of the forest. They sometimes enter houses; the first time I saw one in my chamber, wheeling heavily round and round, I mistook it for a pigeon, thinking that a tame one had escaped from the premises of one of my neighbours. I opened the stomachs of several of these bats, and found them to contain a mass of pulp and seeds of fruits, mingled with a few remains of insects. The natives say they devour ripe cajus and guavas on trees in the gardens, but on comparing the seeds taken from their stomachs with those of all cultivated trees at Ega, I found they were unlike any of them; it is, therefore, probable that they generally resort to the forest to feed, coming to the village in the morning to sleep, because they find it more secure from animals of prey than their natural abides in the woods.

Birds. – I have already had occasion to mention several of the more interesting birds found in the Ega district. The first thing that would strike a newcomer in the forests of the Upper Amazons would be the general scarcity of birds; indeed, it often happened that I did not meet with a single bird during a whole day's ramble in the richest and most varied parts of the woods. Yet the country is tenanted by many hundred species, many of which are, in reality, abundant, and some of them conspicuous from their brilliant plumage. The cause of their apparent rarity is to be sought in the sameness and density of the thousand miles of forest which constitute their dwelling-place. The birds of the country are gregarious, at least during the season when they are most readily found; but the frugivorous kinds are to be met with only when certain wild fruits are ripe, and to know the exact localities of the trees requires months of experience. It would not be supposed that the insectivorous birds are also gregarious, but they are so – numbers of distinct species, belonging to many different families, joining together in the chase or search of food. The proceedings of these associated bands of insect-hunters are not a little curious, and merit a few remarks.

While hunting along the narrow pathways that are made through the forest in the neighbourhood of houses and villages, one may pass several days without seeing many birds; but now and then the surrounding bushes and trees appear suddenly to swarm with them. There are scores, probably hundreds of birds, all moving about with the greatest activity – woodpeckers and Dendrocolaptidae (from species no larger than a sparrow to others the size of a crow) running up the tree trunks; tanagers, ant-thrushes, hummingbirds, fly-catchers, and barbets flitting about the leaves and lower branches. The bustling crowd loses no time, and although moving in concert, each bird is occupied, on its own account, in searching bark or leaf or twig; the barbets visit every clayey nest of termites on the trees which lie in the line of march. In a few minutes the host is gone, and the forest path remains deserted and silent as before. I became, in course of time, so accustomed to this habit of birds in the woods near Ega, that I could generally find the flock of associated marauders whenever I wanted it. There appeared to be only one of these flocks in each small district; and, as it traversed chiefly a limited tract of woods of second growth, I used to try different paths until I came up with it.

The Indians have noticed these miscellaneous hunting parties of birds, but appear not to have observed that they are occupied in searching for insects. They have supplied their want of knowledge, in the usual way of half-civilised people, by a theory which has degenerated into a myth, to the effect that the onward moving bands are led by a little grey bird, called the Uira-para, which fascinates all the rest, and leads them a weary dance through the thickets. There is certainly some appearance of truth in this explanation, for sometimes stray birds encountered in the line of march, are seen to be drawn into the throng, and purely frugivorous birds are now and then found mixed up with the rest, as though led away by some will-o'-the-wisp. The native women, even the white and half-caste inhabitants of the towns, attach a superstitious value to the skin and feathers of the Uira-para, believing that if they keep them in their clothes' chest, the relics will have the effect of attracting for the happy possessors a train of lovers and followers. These birds are consequently in great demand in some places, the hunters selling them at a high price to the foolish girls, who preserve the bodies by drying flesh and feathers together in the sun. I could never get a sight of this famous little bird in the forest. I once employed Indians to obtain specimens for me; but, after the same man (who was a noted woodsman) brought me, at different times, three distinct species of birds as the Uira-para, I gave up the story as a piece of humbug. The simplest explanation appears to be this: the birds associate in flocks from the instinct of self-preservation in order to be a less easy prey to hawks, snakes, and other enemies than they would be if feeding alone.

Toucans – Cuvier's Toucan. – Of this family of birds, so conspicuous from the great size and light structure of their beaks, and so characteristic of tropical American forests, five species inhabit the woods of Ega. The commonest is Cuvier's Toucan, a large bird, distinguished from its nearest relatives by the feathers at the bottom of the back being of a saffron hue instead of red. It is found more or less numerously throughout the year, as it breeds in the neighbourhood, laying its eggs in holes of trees, at a great height from the ground. During most months of the year, it is met with in single individuals or small flocks, and the birds are then very wary. Sometimes one of these little bands of four or five is seen perched, for hours together, among the topmost branches of high trees, giving vent to their remarkably loud, shrill, yelping cries, one bird, mounted higher than the rest, acting, apparently, as leader of the inharmonious chorus; but two of them are often heard yelping alternately, and in different notes. These cries have a vague resemblance to the syllables Tocano, Tocano, and hence, the Indian name of this genus of birds. At these times it is difficult to get a shot at Toucans, for their senses are so sharpened that they descry the hunter before he gets near the tree on which they are perched, although he may be half-concealed among the underwood, 150 feet below them. They stretch their necks downwards to look beneath, and on espying the least movement among the foliage, fly off to the more inaccessible parts of the forest. Solitary Toucans are sometimes met with at the same season, hopping silently up and down the larger boughs, and peering into crevices of the tree-trunks. They moult in the months from March to June, some individuals earlier, others later. This season of enforced quiet being passed, they make their appearance suddenly in the dry forest, near Ega, in large flocks, probably assemblages of birds gathered together from the neighbouring Ygapo forests, which are then flooded and cold. The birds have now become exceedingly tame, and the troops travel with heavy laborious flight from bough to bough among the lower trees. They thus become an easy prey to hunters, and everyone at Ega who can get a gun of any sort and a few charges of powder and shot, or a blow-pipe, goes daily to the woods to kill a few brace for dinner; for, as already observed, the people of Ega live almost exclusively on stewed and roasted Toucans during the months of June and July, the birds being then very fat and the meat exceedingly sweet and tender.

No one, on seeing a Toucan, can help asking what is the use of the enormous bill, which, in some species, attains a length of seven inches, and a width of more than two inches. A few remarks on this subject may be here introduced. The early naturalists, having seen only the bill of a Toucan, which was esteemed as a marvellous production by the virtuosi of the sixteenth and seventeenth centuries, concluded that the bird must have belonged to the aquatic and web-footed order, as this contains so many species of remarkable development of beak, adapted for seizing fish. Some travellers also related fabulous stories of Toucans resorting to the banks of rivers to feed on fish, and these accounts also encouraged the erroneous views of the habits of the birds which for a long time prevailed. Toucans, however, are now well known to be eminently arboreal birds, and to belong to a group including trogons, parrots, and barbets – all of whose members are fruit-eaters. On the Amazons, where these birds are very common, no one pretends ever to have seen a Toucan walking on the ground in its natural state, much less acting the part of a swimming or wading bird. Professor Owen found, on dissection, that the gizzard in Toucans is not so well adapted for the trituration of food as it is in other vegetable feeders, and concluded, therefore, as Broderip had observed the habit of chewing the cud in a tame bird, that the great toothed bill was useful in holding and remasticating the food. The bill can scarcely be said to be a very good contrivance for seizing and crushing small birds, or taking them from their nests in crevices of trees, habits which have been imputed to Toucans by some writers. The hollow, cellular structure of the interior of the bill, its curved and clumsy shape, and the deficiency of force and precision when it is used to seize objects, suggest a want of fitness, if this be the function of the member. But fruit is undoubtedly the chief food of Toucans, and it is in reference to their mode of obtaining it that the use of their uncouth bills is to be sought. Flowers and fruit on the crowns of the large trees of South American forests grow, principally, towards the end of slender twigs, which will not bear any considerable weight; all animals, therefore, which feed upon fruit, or on insects contained in flowers, must, of course, have some means of reaching the ends of the stalks from a distance. Monkeys obtain their food by stretching forth their long arms and, in some instances, their tails, to bring the fruit near to their mouths. Hummingbirds are endowed with highly perfected organs of flight with corresponding muscular development by which they are enabled to sustain themselves on the wing before blossoms whilst rifling them of their contents. These strong-flying creatures, however, will, whenever they can get near enough, remain on their perches while probing neighbouring flowers for insects. Trogons have feeble wings, and a dull, inactive temperament. Their mode of obtaining food is to station themselves quietly on low branches in the gloomy shades of the forest, and eye the fruits on the surrounding trees – darting off, as if with an effort, every time they wish to seize a mouthful, and returning to the same perch. Barbets (Capitoninae) seem to have no especial endowment, either of habits or structure, to enable them to seize fruits; and in this respect they are similar to the Toucans, if we leave the bill out of question, both tribes having heavy bodies, with feeble organs of flight, so that they are disabled from taking their food on the wing. The purpose of the enormous bill here becomes evident; it is to enable the Toucan to reach and devour fruit while remaining seated, and thus to counterbalance the disadvantage which its heavy body and gluttonous appetite would otherwise give it in the competition with allied groups of birds. The relation between the extraordinarily lengthened bill of the Toucan and its mode of obtaining food, is therefore precisely similar to that between the long neck and lips of the Giraffe and the mode of browsing of the animal. The bill of the Toucan can scarcely be considered a very perfectly-formed instrument for the end to which it is applied, as here explained; but nature appears not to invent organs at once for the functions to which they are now adapted, but avails herself, here of one already-existing structure or instinct, there of another, according as they are handy when need for their further modification arises.

One day, while walking along the principal pathway in the woods near Ega, I saw one of these Toucans seated gravely on a low branch close to the road, and had no difficulty in seizing it with my hand. It turned out to be a runaway pet bird; no one, however, came to own it, although I kept it in my house for several months. The bird was in a half-starved and sickly condition, but after a few days of good living it recovered health and spirits, and became one of the most amusing pets imaginable. Many excellent accounts of the habits of tame Toucans have been published, and therefore, I need not describe them in detail, but I do not recollect to have seen any notice of their intelligence and confiding disposition under domestication, in which qualities my pet seemed to be almost equal to parrots. I allowed Tocano to go free about the house, contrary to my usual practice with pet animals, he never, however, mounted my working-table after a smart correction which he received the first time he did it. He used to sleep on the top of a box in a corner of the room, in the usual position of these birds, namely, with the long tail laid right over on the back, and the beak thrust underneath the wing. He ate of everything that we eat; beef, turtle, fish, farinha, fruit, and was a constant attendant at our table – a cloth spread on a mat. His appetite was most ravenous, and his powers of digestion quite wonderful. He got to know the meal hours to a nicety, and we found it very difficult, after the first week or two, to keep him away from the dining-room, where he had become very impudent and troublesome. We tried to shut him out by enclosing him in the backyard, which was separated by a high fence from the street on which our front door opened, but he used to climb the fence and hop round by a long circuit to the dining-room, making his appearance with the greatest punctuality as the meal was placed on the table. He acquired the habit, afterwards, of rambling about the street near our house, and one day he was stolen, so we gave him up for lost.

But two days afterwards he stepped through the open doorway at dinner hour, with his old gait, and sly magpie-like expression, having escaped from the house where he had been guarded by the person who had stolen him, and which was situated at the further end of the village.

The Curl-crested Toucan (Pteroglossus Beauharnaisii). – Of the four smaller Toucans, or Arassaris, found near Ega, the Pteroglossus flavirostris is perhaps the most beautiful in colours, its breast being adorned with broad belts of rich crimson and black; but the most curious species, by far, is the Curl-crested, or Beauharnais Toucan. The feathers on the head of this singular bird are transformed into thin, horny plates, of a lustrous black colour, curled up at the ends, and resembling shavings of steel or ebony-wood: the curly crest being arranged on the crown in the form of a wig. Mr Wallace and I first met with this species, on ascending the Amazons, at the mouth of the Solimoens; from that point it continues as a rather common bird on the terra firma, at least on the south side of the river as far as Fonte Boa, but I did not hear of its being found further to the west. It appears in large flocks in the forests near Ega in May and June, when it has completed its moult. I did not find these bands congregated at fruit-trees, but always wandering through the forest, hopping from branch to branch among the lower trees, and partly concealed among the foliage. None of the Arassaris, to my knowledge, make a yelping noise like that uttered by the larger Toucans (Ramphastos); the notes of the curl-crested species are very singular, resembling the croaking of frogs. I had an amusing adventure one day with these birds. I had shot one from a rather high tree in a dark glen in the forest, and entered the thicket where the bird had fallen to secure my booty. It was only wounded, and on my attempting to seize it, set up a loud scream. In an instant, as if by magic, the shady nook seemed alive with these birds, although there was certainly none visible when I entered the jungle. They descended towards me, hopping from bough to bough, some of them swinging on the loops and cables of woody lianas, and all croaking and fluttering their wings like so many furies. If I had had a long stick in my hand I could have knocked several of them over. After killing the wounded one, I began to prepare for obtaining more specimens and punishing the viragos for their boldness; but the screaming of their companion having ceased, they remounted the trees, and before I could reload, every one of them had disappeared.

Insects. – Upwards of 7000 species of insects were found in the neighbourhood of Ega. I must confine myself in this place to a few remarks on the order Lepidoptera, and on the ants, several kinds of which, found chiefly on the Upper Amazons, exhibit the most extraordinary instincts.

I found about 550 distinct species of butterflies at Ega. Those who know a little of Entomology will be able to form some idea of the riches of the place in this department, when I mention that eighteen species of true Papilio (the swallow-tail genus) were found within ten minutes' walk of my house. No fact could speak more plainly for the surpassing exuberance of the vegetation, the varied nature of the land, the perennial warmth and humidity of the climate. But no description can convey an adequate notion of the beauty and diversity in form and colour of this class of insects in the neighbourhood of Ega. I paid special attention to them, having found that this tribe was better adapted than almost any other group of animals or plants to furnish facts in illustration of the modifications which all species undergo in nature, under changed local conditions. This accidental superiority is owing partly to the simplicity and distinctness of the specific character of the insects, and partly to the facility with which very copious series of specimens can be collected and placed side by side for comparison. The distinctness of the specific characters is due probably to the fact that all the superficial signs of change in the organisation are exaggerated, and made unusually plain by affecting the framework, shape, and colour of the wings, which, as many anatomists believe, are magnified extensions of the skin around the breathing orifices of the thorax of the insects. These expansions are clothed with minute feathers or scales, coloured in regular patterns, which vary in accordance with the slightest change in the conditions to which the species are exposed. It may be said, therefore, that on these expanded membranes Nature writes, as on a tablet, the story of the modifications of species, so truly do all changes of the organisation register themselves thereon. Moreover, the same colour-patterns of the wings generally show, with great regularity, the degrees of blood-relationship of the species. As the laws of Nature must be the same for all beings, the conclusions furnished by this group of insects must be applicable to the whole organic world; therefore, the study of butterflies – creatures selected as the types of airiness and frivolity – instead of being despised, will some day be valued as one of the most important branches of Biological science.

Before proceeding to describe the ants, a few remarks may be made on the singular cases and cocoons woven by the caterpillars of certain moths found at Ega. The first that may be mentioned is one of the most beautiful examples of insect workmanship I ever saw. It is a cocoon, about the size of a sparrow's egg, woven by a caterpillar in broad meshes of either buff or rose-coloured silk, and is frequently seen in the narrow alleys of the forest, suspended from the extreme tip of an outstanding leaf by a strong silken thread five or six inches in length. It forms a very conspicuous object, hanging thus in mid-air. The glossy threads with which it is knitted are stout, and the structure is therefore not liable to be torn by the beaks of insectivorous birds, while its pendulous position makes it doubly secure against their attacks, the apparatus giving way when they peck at it. There is a small orifice at each end of the egg-shaped bag, to admit of the escape of the moth when it changes from the little chrysalis which sleeps tranquilly in its airy cage. The moth is of a dull slatey colour, and belongs to the Lithosiide group of the silk-worm family (Bombycidae). When the caterpillar begins its work, it lets itself down from the tip of the leaf which it has chosen by spinning a thread of silk, the thickness of which it slowly increases as it descends. Having given the proper length to the cord, it proceeds to weave its elegant bag, placing itself in the centre and spinning rings of silk at regular intervals, connecting them at the same time by means of cross threads – so that the whole, when finished, forms a loose web, with quadrangular meshes of nearly equal size throughout. The task occupies about four days: when finished, the enclosed caterpillar becomes sluggish, its skin shrivels and cracks, and there then remains a motionless chrysalis of narrow shape, leaning against the sides of its silken cage.

Many other kinds are found at Ega belonging to the same cocoon-weaving family, some of which differ from the rest in their caterpillars possessing the art of fabricating cases with fragments of wood or leaves, in which they live secure from all enemies while they are feeding and growing. I saw many species of these; some of them knitted together, with fine silken threads, small bits of stick, and so made tubes similar to those of caddice-worms; others (Saccophora) chose leaves for the same purpose, forming with them an elongated bag open at both ends, and having the inside lined with a thick web. The tubes of full-grown caterpillars of Saccophora are two inches in length, and it is at this stage of growth that I have generally seen them. They feed on the leaves of Melastoniae, and as in crawling, the weight of so large a dwelling would be greater than the contained caterpillar could sustain, the insect attaches the case by one or more threads to the leaves or twigs near which it is feeding.

Foraging Ants. – Many confused statements have been published in books of travel, and copied in Natural History works, regarding these ants, which appear to have been confounded with the Sauba, a sketch of whose habits has been given in the first chapter of this work [not included here, but well worth seeking out]. The Sauba is a vegetable feeder, and does not attack other animals; the accounts that have been published regarding carnivorous ants which hunt in vast armies, exciting terror wherever they go, apply only to the Ecitons, or foraging ants, a totally different group of this tribe of insects. The Ecitons are called Tauoca by the Indians, who are always on the look-out for their armies when they traverse the forest, so as to avoid being attacked. I met with ten distinct species of them, nearly all of which have a different system of marching; eight were new to science when I sent them to England. Some are found commonly in every part of the country, and one is peculiar to the open campos of Santarem; but, as nearly all the species are found together at Ega, where the forest swarmed with their armies, I have left an account of the habits of the whole genus for this part of my narrative. The Ecitons resemble, in their habits, the Driver ants of Tropical Africa; but they have no close relationship with them in structure, and indeed belong to quite another sub-group of the ant-tribe.

Like many other ants, the communities of Ecitons are composed, besides males and females, of two classes of workers, a large-headed (worker-major) and a small-headed (worker-minor) class. The large-heads have, in some species, greatly lengthened jaws, the small-heads have jaws always of the ordinary shape; but the two classes are not sharply-defined in structure and function, except in two of the species. There is in all of them a little difference among the workers regarding the size of the head; but in some species this is not sufficient to cause a separation into classes, with division of labour; in others, the jaws are so monstrously lengthened in the worker-majors, that they are incapacitated from taking part in the labours which the worker-minors perform; and again, in others the difference is so great that the distinction of classes becomes complete, one acting the part of soldiers, and the other that of workers. The peculiar feature in the habits of the Eciton genus is their hunting for prey in regular bodies, or armies. It is this which chiefly distinguishes them from the genus of common red stinging-ants, several species of which inhabit England, whose habit is to search for food in the usual irregular manner. All the Ecitons hunt in large organised bodies; but almost every species has its own special manner of hunting.

Eciton rapax. – One of the foragers, Eciton rapax, the giant of its genus, whose worker-majors are half-an-inch in length, hunts in single file through the forest. There is no division into classes amongst its workers, although the difference in size is very great, some being scarcely one-half the length of others. The head and jaws, however, are always of the same shape, and a gradation in size is presented from the largest to the smallest, so that all are able to take part in the common labours of the colony. The chief employment of the species seems to be plundering the nests of a large and defenceless ant of another genus (Formica), whose mangled bodies I have often seen in their possession as they were marching away. The armies of Eciton rapax are never very numerous.

Eciton legionis. – Another species, E. legionis, agrees with E. rapax in having workers not rigidly divisible into two classes; but it is much smaller in size, not differing greatly, in this respect, from our common English red ant (Myrmica rubra), which it also resembles in colour. The Eciton legionis lives in open places, and was seen only on the sandy campos of Santarem. The movement of its hosts were, therefore, much more easy to observe than those of all other kinds, which inhabit solely the densest thickets; its sting and bite, also, were less formidable than those of other species. The armies of E. legionis consist of many thousands of individuals, and move in rather broad columns. They are just as quick to break line, on being disturbed, and attack hurriedly and furiously any intruding object, as the other Ecitons. The species is not a common one, and I seldom had good opportunities to watch its habits. The first time I saw an army was one evening near sunset. The column consisted of two trains of ants, moving in opposite directions; one train empty-handed, the other laden with the mangled remains of insects, chiefly larvae and pupae of other ants. I had no difficulty in tracing the line to the spot from which they were conveying their booty: this was a low thicket; the Ecitons were moving rapidly about a heap of dead leaves; but as the short tropical twilight was deepening rapidly, and I had no wish to be benighted on the lonely campos, I deferred further examination until the next day.

On the following morning, no trace of ants could be found near the place where I had seen them the preceding day, nor were there signs of insects of any description in the thicket, but at the distance of eighty or one hundred yards, I came upon the same army, engaged, evidently, on a razzia of a similar kind to that of the previous evening, but requiring other resources of their instinct, owing to the nature of the wound. They were eagerly occupied on the face of an inclined bank of light earth, in excavating mines, whence, from a depth of eight or ten inches, they were extracting the bodies of a bulky species of ant, of the genus Formica. It was curious to see them crowding around the orifices of the mines, some assisting their comrades to lift out the bodies of the Formicae, and others tearing them in pieces, on account of their weight being too great for a single Eciton – a number of carriers seizing each a fragment, and carrying it off down the slope. On digging into the earth with a small trowel near the entrances of the mines, I found the nests of the Formicae, with grubs and cocoons, which the Ecitons were thus invading, at a depth of about eight inches from the surface. The eager freebooters rushed in as fast as I excavated, and seized the ants in my fingers as I picked them out, so that I had some difficulty in rescuing a few intact for specimens. In digging the numerous mines to get at their prey, the little Ecitons seemed to be divided into parties, one set excavating, and another set carrying away the grains of earth. When the shafts became rather deep, the mining parties had to climb up the sides each time they wished to cast out a pellet of earth; but their work was lightened for them by comrades, who stationed themselves at the mouth of the shaft, and relieved them of their burthens, carrying the particles, with an appearance of foresight which quite staggered me, a sufficient distance from the edge of the hole to prevent them from rolling in again. All the work seemed thus to be performed by intelligent cooperation among the host of eager little creatures, but still there was not a rigid division of labour, for some of them, whose proceedings I watched, acted at one time as carriers of pellets, and at another as miners, and all shortly afterwards assumed the office of conveyors of the spoil.

In about two hours, all the nests of Formicae were rifled, though not completely, of their contents, and I turned towards the army of Ecitons, which were carrying away the mutilated remains. For some distance there were many separate lines of them moving along the slope of the bank – but a short distance off, these all converged, and then formed one close and broad column, which continued for some sixty or seventy yards, and terminated at one of those large termitariums or hillocks of white ants which are constructed of cemented material as hard as stone. The broad and compact column of ants moved up the steep sides of the hillock in a continued stream; many, which had hitherto trotted along empty-handed, now turned to assist their comrades with their heavy loads, and the whole descended into a spacious gallery or mine, opening on the top of the termitarium. I did not try to reach the nest, which I supposed to lie at the bottom of the broad mine, and therefore, in the middle of the base of the stony hillock.

Eciton drepanophora. – The commonest species of foraging ants are the Eciton hamata and E. drepanophora, two kinds which resemble each other so closely that it requires attentive examination to distinguish them; yet their armies never intermingle, although moving in the same woods and often crossing each other's tracks. The two classes of workers look, at first sight, quite distinct, on account of the wonderful amount of difference between the largest individuals of the one, and the smallest of the other. There are dwarfs not more than one-fifth of an inch in length, with small heads and jaws, and giants half an inch in length with monstrously enlarged head and jaws, all belonging to the same brood. There is not, however, a distinct separation of classes, individuals existing which connect together the two extremes. These Ecitons are seen in the pathways of the forest at all places on the banks of the Amazons, travelling in dense columns of countless thousands. One or other of them is sure to be met with in a woodland ramble, and it is to them, probably, that the stories we read in books on South America apply, of ants clearing houses of vermin, although I heard of no instance of their entering houses, their ravages being confined to the thickest parts of the forest.

When the pedestrian falls in with a train of these

ants, the first signal given him is a twittering and restless movement of small flocks of plain-coloured birds (ant-thrushes) in the jungle. If this be disregarded until he advances a few steps farther, he is sure to fall into trouble, and find himself suddenly attacked by numbers of the ferocious little creatures. They swarm up his legs with incredible rapidity, each one driving his pincer-like jaws into his skin, and with the purchase thus obtained, doubling in its tail, and stinging with all its might. There is no course left but to run for it; if he is accompanied by natives they will be sure to give the alarm, crying 'Tauoca!' and scampering at full speed to the other end of the column of ants. The tenacious insects who have secured themselves to his legs then have to be plucked off one by one, a task which is generally not accomplished without pulling them in twain, and leaving heads and jaws sticking in the wounds.

The errand of the vast ant-armies is plunder, as in the case of Eciton legionis; but from their moving always amongst dense thickets, their proceedings are not so easy to observe as in that species. Wherever they move, the whole animal world is set in commotion, and every creature tries to get out of their way. But it is especially the various tribes of wingless insects that have cause for fear, such as heavy-bodied spiders, ants of other species, maggots, caterpillars, larvae of cock-roaches and so forth, all of which live under fallen leaves, or in decaying wood. The Ecitons do not mount very high on trees, and therefore the nestlings of birds are not much incommoded by them. The mode of operation of these armies, which I ascertained only after long-continued observation, is as follows: the main column, from four to six deep, moves forward in a given direction, clearing the ground of all animal matter dead or alive, and throwing off here and there a thinner column to forage for a short time on the flanks of the main army, and re-enter it again after their task is accomplished. If some very rich place be encountered anywhere near the line of march, for example, a mass of rotten wood abounding in insect larvae, a delay takes place, and a very strong force of ants is concentrated upon it. The excited creatures search every cranny and tear in pieces all the large grubs they drag to light. It is curious to see them attack wasps' nests, which are sometimes built on low shrubs. They gnaw away the papery covering to get at the larvae, pupae, and newly-hatched wasps, and cut everything to tatters, regardless of the infuriated owners which are flying about them. In bearing off their spoil in fragments, the pieces are apportioned to the carriers with some degree of regard to fairness of load: the dwarfs taking the smallest pieces, and the strongest fellows with small heads the heaviest portions. Sometimes two ants join together in carrying one piece, but the worker-majors, with their unwieldy and distorted jaws, are incapacitated from taking any part in the labour. The armies never march far on a beaten path, but seem to prefer the entangled thickets where it is seldom possible to follow them. I have traced an army sometimes for half a mile or more, but was never able to find one that had finished its day's course and returned to its hive. Indeed, I never met with a hive; whenever the Ecitons were seen, they were always on the march.

I thought one day, at Villa Nova, that I had come upon a migratory horde of this indefatigable ant. The place was a tract of open ground near the river side, just outside the edge of the forest, and surrounded by rocks and shrubbery. A dense column of Ecitons was seen extending from the rocks on one side of the little haven, traversing the open space, and ascending the opposite declivity. The length of the procession was from sixty to seventy yards, and yet neither van nor rear was visible. All were moving in one and the same direction, except a few individuals on the outside of the column, which were running rearward, trotting along for a short distance, and then turning again to follow the same course as the main body. But these rearward movements were going on continually from one end to the other of the line, and there was every appearance of there being a means of keeping up a common understanding amongst all the members of the army, for the retrograding ants stopped very often for a moment to touch one or other of their onward-moving comrades with their antennae – a proceeding which has been noticed in other ants, and supposed to be their mode of conveying intelligence. When I interfered with the column or abstracted an individual from it, news of the disturbance was very quickly communicated to a distance of several yards towards the rear, and the column at that point commenced retreating. All the small-headed workers carried in their jaws a little cluster of white maggots, which I thought at the time, might be young larvae of their own colony, but afterwards found reason to conclude were the grubs of some other species whose nests they had been plundering, the procession being most likely not a migration, but a column on a marauding expedition.

The position of the large-headed individuals in the marching column was rather curious. There was one of these extraordinary fellows to about a score of the smaller class. None of them carried anything in their mouths, but all trotted along empty-handed and outside the column, at pretty regular intervals from each other, like subaltern officers in a marching regiment of soldiers. It was easy to be tolerably exact in this observation, for their shining white heads made them very conspicuous amongst the rest, bobbing up and down as the column passed over the inequalities of the road. I did not see them change their position, or take any notice of their small-headed comrades marching in the column, and when I disturbed the line, they did not prance forth or show fight so eagerly as the others. These large-headed members of the community have been considered by some authors as a soldier class, like the similarly-armed caste in termites – but I found no proof of this, at least in the present species, as they always seemed to be rather less pugnacious than the worker-minors, and their distorted jaws disabled them from fastening on a plane surface like the skin of an attacking animal. I am inclined, however, to think that they may act, in a less direct way, as protectors of the community, namely, as indigestible morsels to the flocks of ant-thrushes which follow the marching columns of these Ecitons, and are the most formidable enemies of the species. It is possible that the hooked and twisted jaws of the large-headed class may be effective weapons of annoyance when in the gizzards or stomachs of these birds, but I unfortunately omitted to ascertain whether this was really the fact.

The life of these Ecitons is not all work, for I frequently saw them very leisurely employed in a way that looked like recreation. When this happened, the place was always a sunny nook in the forest. The main column of the army and the branch columns, at these times, were in their ordinary relative positions; but, instead of pressing forward eagerly, and plundering right and left, they seemed to have been all smitten with a sudden fit of laziness. Some were walking slowly about, others were brushing their antennae with their forefeet; but the drollest sight was their cleaning one another. Here and there an ant was seen stretching forth first one leg and then another, to be brushed or washed by one or more of its comrades, who performed the task by passing the limb between the jaws and the tongue, and finishing by giving the antennae a friendly wipe. It was a curious spectacle, and one well calculated to increase one's amazement at the similarity between the instinctive actions of ants and the acts of rational beings, a similarity which must have been brought about by two different processes of development of the primary qualities of mind. The actions of these ants looked like simple indulgence in idle amusement. Have these little creatures, then, an excess of energy beyond what is required for labours absolutely necessary to the welfare of their species, and do they thus expend it in mere sportiveness, like young lambs or kittens, or in idle whims like rational beings? It is probable that these hours of relaxation and cleaning may be indispensable to the effective performance of their harder labours, but while looking at them, the conclusion that the ants were engaged merely in play was irresistible.

Eciton praedator. – This is a small dark-reddish species, very similar to the common red stinging-ant of England. It differs from all other Ecitons in its habit of hunting, not in columns, but in dense phalanxes consisting of myriads of individuals, and was first met with at Ega, where it is very common. Nothing in insect movements is more striking than the rapid march of these large and compact bodies. Wherever they pass all the rest of the animal world is thrown into a state of alarm. They stream along the ground and climb to the summits of all the lower trees, searching every leaf to its apex, and whenever they encounter a mass of decaying vegetable matter, where booty is plentiful, they concentrate, like other Ecitons, all their forces upon it, the dense phalanx of shining and quickly-moving bodies, as it spreads over the surface, looking like a flood of dark-red liquid. They soon penetrate every part of the confused heap, and then, gathering together again in marching order, onward they move. All soft-bodied and inactive insects fall an easy prey to them, and, like other Ecitons, they tear their victims in pieces for facility of carriage. A phalanx of this species, when passing over a tract of smooth ground, occupies a space of from four to six square yards; on examining the ants closely they are seen to move, not altogether in one straightforward direction, but in variously spreading contiguous columns, now separating a little from the general mass, now re-uniting with it. The margins of the phalanx spread out at times like a cloud of skirmishers from the flanks of an army. I was never able to find the hive of this species.

Blind Ecitons. – I will now give a short account of the blind species of Eciton. None of the foregoing kinds have eyes of the faceted or compound structure such as are usual in insects, and which ordinary ants (Formica) are furnished with, but all are provided with organs of vision composed each of a single lens. Connecting them with the utterly blind species of the genus, is a very stoutlimbed Eciton, the E. crassicornis, whose eyes are sunk in rather deep sockets. This ant goes on foraging expeditions like the rest of its tribe, and attacks even the nests of other stinging species (Myrmica), but it avoids the light, moving always in concealment under leaves and fallen branches. When its columns have to cross a cleared space, the ants construct a temporary covered way with granules of earth, arched over, and holding together mechanically; under this, the procession passes in secret, the indefatigable creatures repairing their arcade as fast as breaches are made in it.

Next in order comes the Eciton vastator, which has no eyes, although the collapsed sockets are plainly visible; and, lastly, the Eciton erratica, in which both sockets and eyes have disappeared, leaving only a faint ring to mark the place where they are usually situated. The armies of E. vastator and E. erratica move, as far as I could learn, wholly under covered roads – the ants constructing them gradually but rapidly as they advance. The column of foragers pushes forward step by step under the protection of these covered passages, through the thickets, and upon reaching a rotting log, or other promising hunting-ground, pour into the crevices in search of booty. I have traced their arcades, occasionally, for a distance of one or two hundred yards; the grains of earth are taken from the soil over which the column is passing, and are fitted together without cement. It is this last-mentioned feature that distinguishes them from the similar covered roads made by Termites, who use their glutinous saliva to cement the grains together. The blind Ecitons, working in numbers, build up simultaneously the sides of their convex arcades, and contrive, in a surprising manner, to approximate them and fit in the key-stones without letting the loose uncemented structure fall to pieces. There was a very clear division of labour between the two classes of neuters in these blind species. The large-headed class, although not possessing monstrously-lengthened jaws like the worker-majors in E. hamata and E. drepanophora, are rigidly defined in structure from the small-headed class, and act as soldiers, defending the working community (like soldier Termites) against all comers. Whenever I made a breach in one of their covered ways, all the ants underneath were set in commotion, but the worker-minors remained behind to repair the damage, while the large-heads issued forth in a most menacing manner, rearing their heads and snapping their jaws with an expression of the fiercest rage and defiance.

Departure

June 2, 1859 – At length, on the 2nd of June, I left Para, probably forever; embarking in a North American trading-vessel, the Frederick Demming, for New York, the United States route being the quickest as well as the pleasantest way of reaching England. My extensive private collections were divided into three portions and sent by three separate ships, to lessen the risk of loss of the whole. On the evening of the 3rd of June, I took a last view of the glorious forest for which I had so much love, and to explore which I had devoted so many years. The saddest hours I ever recollect to have spent were those of the succeeding night when, the Mameluco pilot having left us free of the shoals and out of sight of land though within the mouth of the river at anchor waiting for the wind, I felt that the last link which connected me with the land of so many pleasing recollections was broken. The Paraenses, who are fully aware of the attractiveness of their country, have an alliterative proverb, 'Quem vai para (o) Para para,' 'He who goes to Para stops there,' and I had often thought I should myself have been added to the list of examples. The desire, however, of seeing again my parents and enjoying once more the rich pleasures of intellectual society, had succeeded in overcoming the attractions of a region which may be fittingly called a Naturalist's Paradise. During this last night on the Para river, a crowd of unusual thoughts occupied my mind. Recollections of English climate, scenery, and modes of life came to me with a vividness I had never before experienced, during the eleven years of my absence. Pictures of startling clearness rose up of the gloomy winters, the long grey twilights, murky atmosphere, elongated shadows, chilly springs, and sloppy summers; of factory chimneys and crowds of grimy operatives, rung to work in early morning by factory bells; of union workhouses, confined rooms, artificial cares, and slavish conventionalities. To live again amidst these dull scenes, I was quitting a country of perpetual summer, where my life had been spent like that of three-fourths of the people – in gipsy fashion – on the endless streams or in the boundless forests. I was leaving the equator, where the well-balanced forces of Nature maintained a land-surface and climate that seemed to be typical of mundane order and beauty, to sail towards the North Pole, where lay my home under crepuscular skies somewhere about fifty-two degrees of latitude. It was natural to feel a little dismayed at the prospect of so great a change; but now, after three years of renewed experience of England, I find how incomparably superior is civilised life, where feelings, tastes, and intellect find abundant nourishment, to the spiritual sterility of half-savage existence, even though it be passed in the garden of Eden. What has struck me powerfully is the immeasurably greater diversity and interest of human character and social conditions in a single civilised nation, than in equatorial South

America, where three distinct races of man live together. The superiority of the bleak north to tropical regions, however, is only in their social aspect, for I hold to the opinion that, although humanity can reach an advanced state of culture only by battling with the inclemencies of nature in high latitudes, it is under the equator alone that the perfect race of the future will attain to complete fruition of man's beautiful heritage, the earth.

The following day, having no wind, we drifted out of the mouth of the Para with the current of fresh water that is poured from the mouth of the river, and in twenty-four hours advanced in this way seventy miles on our road. On the 6th of June, when in 7'55' N. lat. and 52'30' W. long., and therefore about 400 miles from the mouth of the main Amazons, we passed numerous patches of floating grass mingled with tree-trunks and withered foliage. Among these masses I espied many fruits of that peculiarly Amazonian tree the Ubussu palm; this was the last I saw of the Great River.